Katrin Wolff

Protein structure prediction and folding dynamics

Katrin Wolff

Protein structure prediction and folding dynamics

An investigation of protein structure and folding using structural profiles

Südwestdeutscher Verlag für Hochschulschriften

Impressum/Imprint (nur für Deutschland/ only for Germany)
Bibliografische Information der Deutschen Nationalbibliothek: Die Deutsche Nationalbibliothek verzeichnet diese Publikation in der Deutschen Nationalbibliografie; detaillierte bibliografische Daten sind im Internet über http://dnb.d-nb.de abrufbar.

Alle in diesem Buch genannten Marken und Produktnamen unterliegen warenzeichen-, marken- oder patentrechtlichem Schutz bzw. sind Warenzeichen oder eingetragene Warenzeichen der jeweiligen Inhaber. Die Wiedergabe von Marken, Produktnamen, Gebrauchsnamen, Handelsnamen, Warenbezeichnungen u.s.w. in diesem Werk berechtigt auch ohne besondere Kennzeichnung nicht zu der Annahme, dass solche Namen im Sinne der Warenzeichen- und Markenschutzgesetzgebung als frei zu betrachten wären und daher von jedermann benutzt werden dürften.

Verlag: Südwestdeutscher Verlag für Hochschulschriften Aktiengesellschaft & Co. KG
Dudweiler Landstr. 99, 66123 Saarbrücken, Deutschland
Telefon +49 681 37 20 271-1, Telefax +49 681 37 20 271-0
Email: info@svh-verlag.de
Zugl.: Darmstadt, TU, Diss., 2010

Herstellung in Deutschland:
Schaltungsdienst Lange o.H.G., Berlin
Books on Demand GmbH, Norderstedt
Reha GmbH, Saarbrücken
Amazon Distribution GmbH, Leipzig
ISBN: 978-3-8381-1520-7

Imprint (only for USA, GB)
Bibliographic information published by the Deutsche Nationalbibliothek: The Deutsche Nationalbibliothek lists this publication in the Deutsche Nationalbibliografie; detailed bibliographic data are available in the Internet at http://dnb.d-nb.de.

Any brand names and product names mentioned in this book are subject to trademark, brand or patent protection and are trademarks or registered trademarks of their respective holders. The use of brand names, product names, common names, trade names, product descriptions etc. even without a particular marking in this works is in no way to be construed to mean that such names may be regarded as unrestricted in respect of trademark and brand protection legislation and could thus be used by anyone.

Publisher: Südwestdeutscher Verlag für Hochschulschriften Aktiengesellschaft & Co. KG
Dudweiler Landstr. 99, 66123 Saarbrücken, Germany
Phone +49 681 37 20 271-1, Fax +49 681 37 20 271-0
Email: info@svh-verlag.de

Printed in the U.S.A.
Printed in the U.K. by (see last page)
ISBN: 978-3-8381-1520-7

Copyright © 2010 by the author and Südwestdeutscher Verlag für Hochschulschriften Aktiengesellschaft & Co. KG and licensors
All rights reserved. Saarbrücken 2010

Contents

1 Introduction   **1**
  1.1 Motivation 1
  1.2 Protein Structure 3
  1.3 Contact Maps and Structural Profiles 6

2 Protein Structure Prediction   **11**
  2.1 Motivation 11
  2.2 Structure Selection 12
      2.2.1 Benchmark Structures 13
      2.2.2 Prediction of Structural Profiles 16
      2.2.3 Distribution of Rosetta Candidate Structures 17
      2.2.4 Filtering of Rosetta Candidates 19
      2.2.5 Filtering of CASP8 Models 27
  2.3 Discussion 28

3 Protein Folding Dynamics   **31**
  3.1 Motivation 31
  3.2 The Protein Model 32
  3.3 Protein Structure Reconstruction 39
  3.4 Folding Simulations and Free Energy Landscapes 41
      3.4.1 Folding Simulations, Time Series and Distributions of Observables 42
      3.4.2 Constrained Sampling 46
      3.4.3 Metadynamics 48
      3.4.4 Constrained Sampling and Metadynamics Combined 52
      3.4.5 Example Proteins and Comparison of Models 53
      3.4.6 Contact Maps as Microstates 65
      3.4.7 Heat Capacities and Folding Transitions 67
  3.5 Discussion 70

4 Conclusion and Outlook   **73**

A RMSD and TM-Score Distribution of Candidate Structures   **75**

List of Figures

1.1 Amino acids and peptide bond 3
1.2 Protein sequence and protein structure 5
1.3 Coarse-grained protein structure 6
1.4 Contact map and structure profiles 7

2.1 Artificial Neural Network (ANN) to predict structural profile from sequence 16
2.2 Distribution of RMSD and 1−TM-score for proteins 1pv0 and 1ubq 18
2.3 Correlation of the Rosetta score and the EC-score for exact profile and predicted profile, to RMSD for proteins 1ubq and 1shg 21
2.4 RMSD and 1−TM-score distribution of filtered structures for proteins 1c9oA and 1ubq 22
2.5 RMSD distribution of filtered and clustered structures for proteins 1c9oA and 1btb 23
2.6 Number of good structures in dependance of number of selected structures for proteins 1c9oA and 1ubq 23
2.7 Distribution of filtered structures for interpolated ECs for proteins 1gb1 and 1c9oA 26
2.8 Relative frequencies of good structures for the CASP8 set 28

3.1 Tube model 33
3.2 Move set of Monte Carlo simulation 33
3.3 Restricted structural profiles 36
3.4 Geometric criterion of secondary structure 37
3.5 Statistics of reconstructed proteins 39
3.6 Examples of reconstructed structures 40
3.7 Distribution of full contact overlap 41
3.8 Folding time series for the villin headpiece in the EC-model 43
3.9 Folding time series for the villin headpiece in the Gō-model 43
3.10 Definition of parts A and B for the villin headpiece 44
3.11 Folding of the villin headpiece in partial RMSD coordinates 45
3.12 Folding of the villin headpiece in helix content and contact number coordinates . 46
3.13 Constrained sampling for the villin headpiece 47
3.14 Four state system 49
3.15 Metadynamics sampling for the villin headpiece 51
3.16 Constrained metadynamics sampling for the villin headpiece 52
3.17 Combined metadynamics samplings for the villin headpiece 53
3.18 Villin headpiece target structure 54
3.19 Free energy landscape for the villin headpiece in the EC- and Gō-model in helix content and number of contacts 55
3.20 Free energy landscape for the villin headpiece in the EC-model in $RMSD_A$ and $RMSD_B$ 56
3.21 Free energy landscape for the villin headpiece in the Gō-model in $RMSD_A$ and $RMSD_B$ 57
3.22 Free energy landscape for the villin headpiece with folding trajectories 58

3.23 BBL target structure 60
3.24 Free energy profile for BBL in the EC-model in end-to-end distance 60
3.25 Free energy profile for BBL in the EC-model in contact overlap 61
3.26 Free energy profile for BBL in the Gō-model in contact overlap 61
3.27 WW domain target structure 62
3.28 Free energy landscape for the WW domain in the EC- and Gō-model in helix content and contact number 63
3.29 Folding time series for the WW domain in the EC-model for energy and helix content 64
3.30 Free energy landscape for the WW domain in the EC-model in contact overlap and RMSD 64
3.31 Free energy landscape for the WW domain in the Gō-model in contact overlap and RMSD 65
3.32 Frequent contacts and frequent contact maps 67
3.33 Heat capacity curves for the villin headpiece in the EC- and Gō-model 68
3.34 Energy distribution for the villin headpiece in the EC-model 69
3.35 Energy distribution for the villin headpiece in the Gō-model 69
3.36 Configurations of the villin headpiece in the EC-model at different temperatures . 70

A.1 Distribution of RMSD and 1−TM-score for proteins 1pv0, 1gb1, 1shg and 1jic. . . 77
A.1 Distribution of RMSD and 1−TM-score for proteins 1r69, 1c9oA, 1mjc and 1fgp . 78
A.1 Distribution of RMSD and 1−TM-score for proteins 1ubq, 1oqp, 1btb and 1p9yA . 79
A.1 Distribution of RMSD and 1−TM-score for proteins 2imf, 1volA, 1ix9A and 1f5x . 80
A.1 Distribution of RMSD and 1−TM-score for proteins 1gk9A and 1by1 81

List of Tables

1.1	Canonical proteinogenic amino acids	4
2.1	Protein target structures for which candidates were predicted using Rosetta	14
2.2	Protein target structures from CASP8	15
2.3	Quality of different filtering methods	24
2.4	Filtering by interpolated structural profiles	25
2.5	Centres of the 10 largest clusters	27
3.1	Free energy differences for four state system	49
A.1	Estimated number of candidates to expect one structure below RMSD=5 Å for large protein target structures	75

1 Introduction

1.1 Motivation

Proteins carry out a variety of vital functions in every living organism, ranging from enzyme proteins regulating biochemical reactions to motor proteins causing the contraction and movement of muscles. For the vast majority of proteins the three-dimensional shape is crucial for biological function [1].

While the amino acid sequence constituting the protein defines the three-dimensional form at physiological conditions (usually uniquely) [2] it is not at all clear how this mapping from sequence to structure should be performed. Experimental structure determination by X-ray or NMR studies on the other hand is very costly compared to the relatively simple sequencing. Therefore, there are a host of protein sequences available while only a much smaller number of structures are known [3, 4]. Thus, prediction of protein structure from sequence poses a formidable challenge with great impact on protein engineering and medical applications.

Not only the final biologically active form, the so-called native state of a protein, is of scientific interest but also the folding process that may take microseconds for the fastest folders or up to several minutes for more complex proteins [1]. Proteins in living organisms occasionally fold into "wrong" shapes that cannot fulfill the biological function and, although these misfolded proteins are usually quickly degraded, they are connected to diseases such as BSE or Alzheimer's [1].

The two parts, protein structure prediction and folding dynamics, thus approach the problem of how a chain of amino acids folds into a specific structure from two different angles. While the former focuses on the final structure, i.e. the outcome of folding, the latter elucidates the physical process itself. Successful methods of structure prediction usually ignore the physical process, whereas investigations on folding behaviour often incorporate information on the native structure. Although these methods differ, both cases can gain from the use of structural profiles, as will be shown in this thesis.

Protein structures can be represented as so-called structural profiles [5, 6] (see Section 1.3). Such a profile is an array of the same length as the protein's amino acid sequence and contains information on each position's connectivity, or propensity to have contacts with other parts of the sequence. As such it is strongly correlated to the hydrophobicity of the amino acid at the position in question [7].

Prediction of the unknown structure of a protein can principally follow one of two routes: Either the sequence is found to be highly similar to that of one or more proteins of known structure, then a template can be created from the known structure(s) and used as a basis for structure prediction. This method is known as homology modelling which today gives very acceptable results [8]. If this is not the case and no sufficiently similar structure can be found, the structure has to be predicted *ab initio* and only local similarities to known structures can be exploited. A common method then is to proceed in two steps, the first of which consists in creating a large number of coarse-grained candidate structures. The second step is to refine these first guesses to a higher level of detail. As refinement is computationally very costly, it is very important to limit this task to the most promising candidates. The first part of this thesis, Chapter 2, addresses this problem of candidate identification and selection.

In this context of *ab initio* structure prediction the use of the exact profile derived from the structure will be discussed as a proof of principle for the selection of good candidate structures. As these exact profiles, however, are not available for real predictions, the true challenge is to use profiles in the selection step that have been predicted from sequence. Filtering by either exact or predicted profiles is tested on two different sets of proteins and compared to the more established methods of filtering by low-resolution energy [9] and structure selection by clustering [10, 11]. The influence of profile prediction quality on the performance of filtering is investigated, as are size of proteins and quality of candidate sets.

Simulations of protein folding can give insight into intermediate structures or long-lived metastable states as well as elucidate the folding mechanism. Such simulations range from highly simplified models on two- or three-dimensional lattices [12, 13], beads-on-a-string models with only hydrophobic interactions [14, 15] or Gō-models based on the principle of minimum frustration [12, 16] to molecular dynamics (MD) simulations of varying complexity. For a protein in an aqueous solution the effects of water can be taken into account implicitly via solvation models (implicit water, e.g. Ref. [17]). The more exact approach is to simulate a few layers of molecular water along with the protein itself (explicit water) [18]. This, however, dramatically increases computational costs as it not only increases the system size but at the same time decreases the possible integration time step. Likewise, an all-atom force field [19] or an effective force field with so-called united residues [20, 21] can be used. Of course these methods all have different advantages and short-comings. While lattice proteins in their discrete conformation space can be exhaustively enumerated, and are especially useful if generic folding behaviour is investigated [13, 22], they are not realistic enough to study specific proteins. Highly sophisticated MD force fields on the other hand can capture a high level of details close to a protein's native state but are not well-suited to investigate entire folding trajectories. This is due to the high computational cost and, more importantly, to today's force fields, which are optimised for fully folded structures and small molecules only [23] so that their behaviour is not necessarily realistic for unfolded conformations. There are also a number of purely theoretical models [24–26] that make predictions on heat capacity curves and the like but not on folding pathways.

In the second part of this thesis, Chapter 3, I study the dynamics of proteins by means of Monte Carlo simulations in a coarse-grained model biased towards the native structure. This approach is similar in spirit to Gō-models which also rely on the knowlegde of the native state but it differs in so far as it does not assume that only those amino acids that interact in the native state interact during folding (i.e. the principle of minimal frustration). In the context of folding dynamics the structural profile is computed from the known three-dimensional structure and used to create a potential favouring the native structure. This thesis shows that the model based on the structural profile allows the successful reconstruction of three-dimensional protein structures, and investigates the folding behaviour by means of folding trajectories and free energy landscapes. Adapted sampling schemes to create free energy landscapes are presented and discussed. Some specific example proteins, for which experimental data or molecular dynamics simulations are available, are examined in more detail. These results are compared to those obtained by Gō-models and it is shown that by adding only a little more complexity to the model considerably more realistic behaviour can be observed.

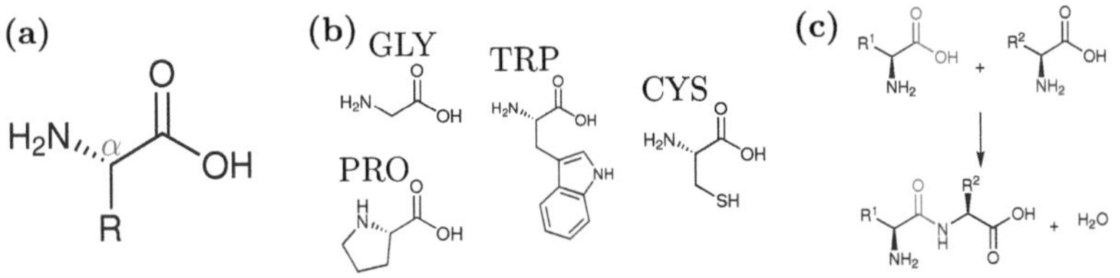

Figure 1.1: Amino acids and peptide bond. Part **(a)** shows a generic amino acid with R denoting the side chain that distinguishes the amino acid, the red α indicates the position of the α-carbon or C_α. Some specific amino acids are given in **(b)**, glycine (GLY) is the simplest amino acid where the entire side chain consists of a single hydrogen atom, whereas tryptophan (TRP) is particularly bulky and aromatic. Proline (PRO) is special in that its side chain closes back in on the nitrogen atom and cysteine (CYS) can form very stable disulphide bonds with other cysteines in the protein. Part **(c)** shows the formation of a peptide bond between two amino acids. For those amino acids where the C_α-atom is a chiral centre the naturally abundant enantiomer is given. Drawings of chemical structures are courtesy of C. Wolff.

1.2 Protein Structure

Naturally occurring proteins mostly consist of 20 standard (so-called canonical) amino acids as building blocks. These amino acids form linear (unbranched) chains or sequences which subsequently fold into highly specific three-dimensional structures. Each amino acid has an amino(N)-terminus and a carboxyl(C)-terminus (see Fig. 1.1 (a)) and a side chain R ("residue" or simply "rest") that determines the amino acid's identity. The carbon atom to which the side chain is attached is the C_α-atom. A list of canonical amino acids is given in Table 1.1 where they are loosely grouped according to their polarity and charge, following Ref. [27]. These properties are important to account for an amino acid's hydrophobicity or hydrophilicity and therefore its tendency to be part of the protein core or part of the solvent-exposed surface.

Another important property is sheer size – while small side chains (such as the single hydrogen atom of glycine, see Fig. 1.1 **(b)**) can be packed into tight turns, large side chains (e.g. that of tryptophan) are too bulky. There are two amino acids that can be marked as special for their structural properties: Proline's side chain closes back in on its backbone's nitrogen atom making proline particularly rigid. As such proline frequently disrupts secondary structural elements. Cysteine on the other hand even has impact on tertiary (or quaternary) structure by forming very stable disulphide bonds with other cysteines that can be far away in the sequence or even located on another chain within a protein complex.

In protein sequences amino acids are usually abbreviated to either a three- or a one-letter code. These abbreviations are also reported in Table 1.1. There are more abbreviations for unknown or unclear amino acids (e.g. if the amino acid is leucine or isoleucine but the exact kind could not be determined this is indicated by the three-letter code XLE and one-letter code J). One relatively frequent non-standard amino acid is selenomethionine (MSE) which is incorporated by the organism instead of methionine indiscriminately, the sole difference between the two being that selenomethionine contains a selenium atom instead of the sulphur atom.

amino acid	three-letter code	one-letter code	side-chain polarity
glycine	GLY	G	nonpolar
alanine	ALA	A	nonpolar
valine	VAL	V	nonpolar
leucine	LEU	L	nonpolar
isoleucine	ILE	I	nonpolar
phenylalanine	PHE	F	nonpolar
proline	PRO	P	nonpolar
methionine	MET	M	nonpolar
tryptophan	TRP	W	nonpolar
aspartic acid	ASP	D	charged
glutamic acid	GLU	E	charged
lysine	LYS	K	charged
arginine	ARG	R	charged
serine	SER	S	polar
threonine	THR	T	polar
tyrosine	TYR	Y	polar
histidine	HIS	H	polar
cysteine	CYS	C	polar
asparagine	ASN	N	polar
glutamine	GLN	Q	polar

Table 1.1: The 20 canonical proteinogenic amino acids, the first three columns contain full name and three- and one-letter codes. The last column groups amino acids into three classes: Nonpolar, charged and polar, respectively. Source: Ref. [27].

Because of this property it is sometimes substituted for methionine and used in X-ray structure determination as its larger mass helps with the crystallography phase problem [28].

The amino acids in a protein's sequence, or primary structure, form peptide bonds (see Fig. 1.1 **(c)** and Fig. 1.2 **(a)**): One amino acid's NH_2-group reacts with an other's COOH-group, the two amino acids become linked and one molecule of water is released. In the rigid peptide bond, consecutive C_α-, C-, N- and C_α-atoms are restrained to lie in a plane with a fixed distance between C_α-atoms of 3.8 Å in the predominant *trans* configuration (2.8 Å in the much rarer *cis* configuration). This is equivalent to the statement that the dihedral angle ω between planes defined by C_α-C-N and C-N-C_α is restricted to values of 180° (*trans*) or 0° (*cis*). The dihedral angles between the plane defined by C-N-C_α and the plane defined by N-C_α-C is called Φ and the dihedral angle between the plane defined by N-C_α-C and the plane defined by C_α-C-N Ψ. These angles, Φ and Ψ, are not restricted by the peptide bond but make up the backbone's degrees of freedom.

To form a compact folded structure and squeeze out water from the interior, proteins form secondary structure elements [29] such as α-helices and β-sheets (see Fig. 1.2 **(b)**). This results in values of Φ and Ψ clustering around typical regions in so-called Ramachandran plots [30] corresponding to α-helices or β-sheets. These secondary structure elements are additionally stabilised by hydrogen bonds between strands of β-sheets or turns in helices [31]. Chirality of amino acids (see Fig. 1.1 **(b)**) results in a preferred chirality of α-helices but the cause of their homochirality (as well as that of the ribose of RNA) is still puzzling [32]. Secondary structure el-

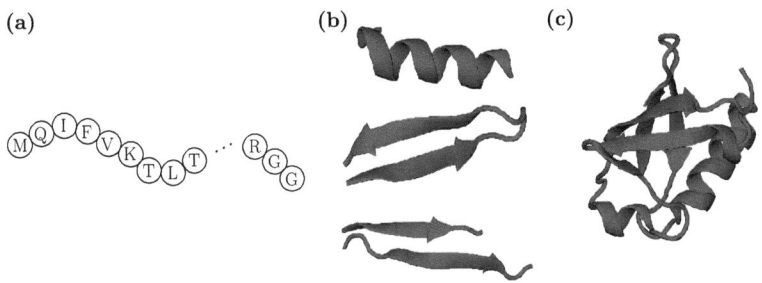

Figure 1.2: Protein sequence and protein structure illustrated on ubiquitin (PDB [35] id. 1ubq). **(a)** Amino acid sequence or primary structure, **(b)** secondary structure elements – helix, anti-parallel and parallel β-sheet – **(c)** and tertiary structure.

ements are then assembled into tertiary structures (see Fig. 1.2 **(c)**) often with recurring motifs such as β-α-β (not shown). Quite often the biologically functional unit consists of several proteins forming complexes. These protein assemblies are then referred to as quaternary structures. They can either contain repeating identical subunits or heterogenous substructures. Although structure determination proceeds at a much slower pace than sequencing of proteins, the first sequence (of insulin [33], 1955) and structure (of myoglobin [34], 1958) were determined at roughly the same time.

A protein's sequence thus determines the native three-dimensional structure which, according to Anfinsen's famous paradigm [2], lies in the free energy minimum and is thus stable at physiological conditions. Specifically, this means that a protein will refold to its native state after denaturation, once physiological conditions are restored. This paradigm holds at least for small proteins. Larger and more complex poteins may require assistance of chaperones for folding [27] and a couple of proteins have been observed for which the biologically active (native) state is not the most stable conformation. These proteins degrade after minutes to hours and become inactive [36–38]. Folding is also influenced by confinement and molecular crowding [39, 40]. Notwithstanding these few (but notable) exceptions, Anfinsen's paradigm is usually assumed in protein folding which is also the case for this thesis. The structures discussed for folding in this thesis are rather small (up to 45 amino acids) and domains of larger proteins. They fold, however, autonomously if excised from the large protein and independently from other domains if part of the large proteins which themselves fold in a modular fashion [1].

In 1968 Levinthal noted that even a protein of moderate size would require astronomic times for folding by a random search of conformations, whereas in fact proteins may fold very quickly. He resolved what seemed to be a paradox by postulating pathways of folding that were followed by the protein [41, 42]. Today this view has been replaced by the notion of a funnel in energy leading towards the native structure [16, 43], thus no clear-cut pathway or sequence of intermediate conformations has to be adhered to but instead multiple routes are allowed that finally reach the bottom of the funnel. While potential energy is funnelled towards the native states (with some possible roughness), loss of entropy almost compensates for the gain in energy during folding and free energy has to overcome a (at least one) barrier. The possibility of downhill folding without a free energy barrier exists but its observation for a real

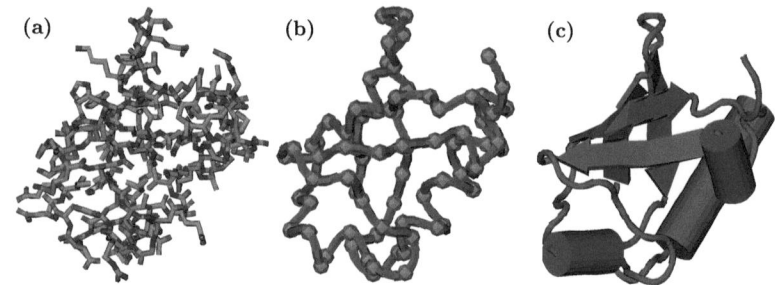

Figure 1.3: Different levels of coarse-graining illustrated on ubiquitin (PDB id. 1ubq), **(a)** shows the positions of all non-hydrogen atoms and covalent bonds between them (carbon is shown in cyan, oxygen in red, nitrogen in blue and sulphur in yellow), **(b)** gives the backbone structure with C_α-atoms highlighted as cyan balls and peptide bonds shown in red, **(c)** takes coarse-graining to the level of secondary structure elements.

protein is currently intensely debated [44–46] (see also the discussion of example protein BBL in section 3.4.5). Levinthal's paradox, however, is still of importance when designing potentials, mainly for structure prediction. Conformation space is too vast to be exhaustively sampled and energy potentials have to be designed to be funnelled towards the native structure if near-native structures are to be found within reasonable times.

Proteins are classified into folds and families based on their structures [47, 48] and it appears that the number of possible folds is limited. New folds for single-domain proteins are rarely encountered and new territory is mostly due to new assemblies of different domains [49]. From this follows that many sequences map onto very similar structures, meaning that structure information is evolutionarily stronger conserved than sequence information [50]. There also exist exceptions from this rule and proteins could be designed to more than 80% sequence identity but folding into very different structures [51] and naturally occurring proteins of different folds but 40% sequence identity were observed [52]. Usually, however, sequence similarity infers even stronger structural similarity and a good way to predict structures is by searching for, even remotely, similar sequences of known structure [8].

1.3 Contact Maps and Structural Profiles

Protein structures can be simplified to different levels of coarse-graining (see Fig. 1.3). The very first step of coarse-graining is to not treat proteins in a quantum chemical description, ignore electron correlations and instead employ interatomic potentials [27] (Fig. 1.3 **(a)**). This will result in the empirical potentials used in all-atom force fields such as CHARMM [53], AMBER [54] or GROMACS [55], to name a few. The next level of simplification usually is to omit the side chains that are specific for every amino acid and consider only the backbone consisting of repeating [NCCO]-units (Fig. 1.3 **(b)**). The inverse step, to include side chains into a given backbone or recover the full information, involves optimising rotamer positions from a side chain library [56] and, although complicated, can be viewed as basically solved.

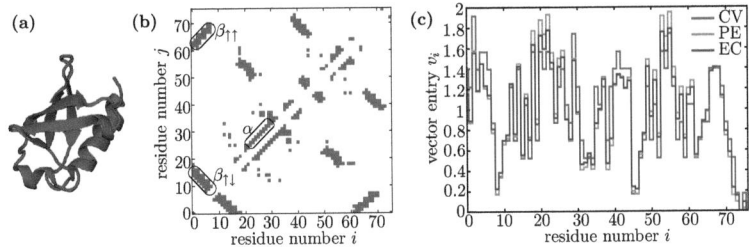

Figure 1.4: From the three-dimensional structure **(a)** the contact map **(b)** can be computed. Possible structure profiles **(c)** are contact vector (CV), principal eigenvector (PE) and effective connectivity (EC). The example protein shown is ubiquitin (PDB id. 1ubq).

Further simplifying the backbone results in a C_α-trace. Because of the rigid peptide bond this representation is essentially equivalent to the backbone of [NCCO]-units and the full all-atom description can be similarly recovered [57]. The next level of simplification would consider elements of secondary structure as basic units [58] (see Fig. 1.3 **(c)**), however, in this thesis we will stay at the abstraction level of representative C_α-atoms (with one exception in Section 3.4.7). Working with secondary structure elements as building blocks also makes the assumption that they are significantly more stable than tertiary structure and precede the latter in formation, which is one theory of folding mechanisms but not the only one. It is therefore more informative to omit this *a priori* assumption and verify afterwards whether folding simulations follow this route.

The distance between consecutive C_α-atoms is (approximately) fixed, a consequence of the rigid peptide bond, which leaves roughly $2L$ degrees of freedom where L is the number of amino acids. (The first amino acid can be placed arbitrarily, the second amino acid arbitrarily on a sphere of radius fixed by the interatomic distance. For the next amino acid one angle has to be specified and for the remaining $L - 3$ amino acids two angles.) A huge simplification is to discard the individual amino acid positions and only retain information about contacts, i.e. close proximity, between amino acids which results in an $L \times L$ symmetric and *binary* matrix, the so-called contact map. In general, this representation is not equivalent to the coordinate information – a chain with no contacts at all can still take a large number of conformations which would all be mapped to the empty matrix. However, for compact folded proteins the coordinate representation can be retrieved from the contact map [59, 60] with only a marginal loss in resolution that is comparable to resolution obtained in experiments. Another small issue that has to be considered is that chirality matters in biological molecules and the mapping to contact matrices does not preserve this property. As helices hardly ever occur in the left-handed version, however, it is easy to pick the correct structure.

There are several ways to define a contact between two amino acids, the simplest being by the distance between C_α-atoms. The distance threshold, or contact radius r_c, is also somewhat arbitrary and whether a definition is a good choice also depends on the intended application. In this thesis, a contact radius of $r_c = 8.5\,\text{Å}$ was found to work well in the context of both structure prediction and folding dynamics. The contact map is then defined as

$$C_{ij} = \begin{cases} 1 & d_{i,j} < r_c \land |i-j| > 2 \\ 0 & d_{i,j} \geq r_c \lor |i-j| \leq 2 \end{cases} \quad (1.1)$$

where $d_{i,j}$ is the distance between C_α-atom i and C_α-atom j. Trivial contacts, such as self-contacts and any contact with $|i-j| \leq 2$, are disregarded. In a similar approach, contacts between amino acids could be defined based on distances between "heavy atoms", which means any non-hydrogen atoms. The distance threshold is then usually set to $r_c = 4.5\,\text{Å}$ and $C_{ij} = 1$ if any heavy atom of amino acid i comes closer than $4.5\,\text{Å}$ to any heavy atom of amino acid j (with analogous treatment of trivial contacts). In the following, however, the three-dimensional structure will be described at the level of C_α-atoms and contacts are defined accordingly.

For structure prediction purposes contact maps are not very useful as they are themselves very difficult to predict [61]. In particular false positives in the prediction of the contact map severely worsen the resulting three-dimensional structure while missing contacts are not as harmful [62].

In the context of folding simulations, use of the contact map as a bias towards the target structure corresponds to Gō-models where native contacts, i.e. interactions in the native state, are made attractive.

Structure information can be further compressed by making use of structural profiles. There is a host of different definitions [6] of one-dimensional representations that convey information about each amino acid's connectivity. A simple structural profile is the contact vector that gives the number of contacts for each amino acid i,

$$\tilde{c}_i = \mathcal{N} \sum_{j=1}^{L} C_{ij}. \quad (1.2)$$

The normalising constant $\mathcal{N}$ can be chosen such that

$$\langle \tilde{c} \rangle = \frac{1}{L} \sum_{i=1}^{L} \tilde{c}_i = 1. \quad (1.3)$$

This choice of profile is useful for structure comparison and alignment since its computation requires very little time [63]. A drawback when it comes to structure prediction or folding simulations is that the contact vector is degenerate when compared to the contact map. Multiple structures that can be quite distinct, in particular only partly folded structures, are mapped onto the same contact vector [64].

Structural profiles that are derived from the contact map's eigensystem are better suited for the task of structure prediction and folding investigations although taking more computing time. The effective connectivity (EC) is the profile of choice in this thesis and contains contributions from all the contact map's eigenvectors weighted according to the corresponding eigenvalues,

$$\mathbf{c} = \frac{1}{A} \sum_{j=1}^{L} \frac{1}{\Lambda - \lambda^{(j)}} \mathbf{v}^{(j)} \langle v^{(j)} \rangle. \quad (1.4)$$

Here $\mathbf{c}$ is the effective connectivity, a vectorial quantity, and the $\mathbf{v}^{(j)}$ ($j = 1, \ldots, L$) are the L eigenvectors of the contact map with their eigenvalues $\lambda^{(j)}$. The quantity $\langle v^{(j)} \rangle$ is the average of entries $v_i^{(j)}$ of eigenvector j, A and Λ are parameters used to fix the average of $\mathbf{c}$, $\langle c \rangle = 1$, and the relative variance $\langle c^2 \rangle / \langle c \rangle^2 = \langle \tilde{c}^2 \rangle / \langle \tilde{c} \rangle^2$ to the same value as that of the contact vector $\tilde{\mathbf{c}}$ [6]. As it turns out, for a single domain structure Λ is often close to the largest eigenvalue, $\Lambda \approx \lambda^{(1)}$, meaning that the contribution of the eigenvector corresponding to the largest eigenvalue will be dominant. Consequently, the eigenvector to the largest eigenvalue, the principal eigenvector (PE), on its own is also a valid choice for a structural profile for single domain structures where $\Lambda \approx \lambda^{(1)}$ and has been used at early stages of this thesis.

Both profile definitions, EC and PE, can be unified [6] as they maximise the quadratic form

$$Q \equiv \sum_{ij} C_{ij} c_i c_j \qquad (1.5)$$

for a given contact map C_{ij} under different side conditions. The well-known definition of the PE is to maximise Q under the constraint that

$$\langle c^2 \rangle = \frac{1}{L} \sum_{i=1}^{L} c_i^2 = 1. \qquad (1.6)$$

For the EC an additional condition is introduced, namely

$$\langle c \rangle = \frac{1}{L} \sum_{i=1}^{L} c_i = 1 \qquad (1.7)$$

and the condition in Eq. (1.6) is changed to $\langle c^2 \rangle / \langle c \rangle^2 = \langle \tilde{c}^2 \rangle / \langle \tilde{c} \rangle^2$ as mentioned above. Maximising Q under these constraints then leads to the expression in Eq. (1.4) where the open parameters A and Λ have to be fixed to satisfy the constraints [6].

From this unifying definition the correlation to hydrophobicity becomes evident. Maximising Q in expression (1.5) means that entries c_i will be large for amino acids i with many contacts to other amino acids j – and even more so if these amino acids have many contacts and large entries c_j of their own. In order to have many contacts, an amino acid will be buried in the protein core, as hydrophobic amino acids tend to do. Moreover, they will have contacts predominantly with other buried, i.e. hydrophobic, amino acids, as again is characteristic of amino acids of high hydrophobicity. An entry c_i of either EC or PE thus depends not only on the number of contacts of amino acid i but also on the contacts of those amino acids with which i is in contact (and the amino acids with which those are in contact etc.) and therefore contains information on amino acid connectivity which is more detailed than that in the contact vector.

There is no mathematical proof that either EC or PE are indeed equivalent to the contact map (under appropriate constraints on the contact map such as connectedness or possibly existence of secondary structure motifs) but for the PE there is a reconstruction algorithm [65] that was found to work for all compact single-domain proteins investigated. For the EC no degeneracy (one EC profile corresponding to multiple contact maps) was encountered in any of the simulations run for this thesis. A drawback of the PE is that, for multi-domain proteins where the contact maps decompose into disconnected blocks, it will only give information about the largest and best-connected block. As the EC contains contributions from all the contact map's eigenvectors, and thus information on all protein domains, it does not display this problem and was consequently used as the structure profile of choice. Although the small proteins investigated in the folding dynamics all consist of a single domain in their native state, the conformations encountered during folding can be more complicated.

These vectorial representations of protein structure as profiles are much more amenable to prediction from sequence than contact maps because they are of the same dimension as the sequence, which is a string of amino acids. The profile's correlation to amino acid hydrophobicity can then be exploited to predict it from a given sequence.

In the context of protein structure prediction, in particular in the selection of structure candidates, predicted profiles can be used. The profile predicted for a given sequence serves as a target to which the profiles calculated from the candidates' structures are compared using a score based on the difference between the two profiles. Those conformations that are in good agreement with the predicted profile are retained for further refinement.

Likewise, for protein folding an energy is defined based on a measure of the difference between the target structure's profile and that of the current conformation. The energy is minimal if the correct profile and thus the correct structure has been reached. As is the case with Gō-models the model thus contains an obvious bias towards the native structure but instead of making only native interactions attractive the energy in the profile-based model relates how well all the amino acids' connectivity or hydrophobicity is satisfied in the current conformation. This results, in effect, in an interaction of all amino acids with all the other amino acids.

2 Protein Structure Prediction

2.1 Motivation

As experimental determination of protein structures by either X-ray crystallography or NMR is very costly, prediction of structure from sequence is of great interest. This prediction can either be based on already known sequence-structure pairs or be performed *ab initio*.

If sequences of high similarity (and known structure) can be found, the method of choice is homology modelling where the known structures are used to create a template on which the structure to be predicted is modelled. In biology, the term "homology" is used to state common ancestry of proteins. In the context used here, however, it is not necessary to establish true homology in the above sense for the selection of sequences. Instead, sequences of known structure are selected based on their similarity to the query sequence which makes homology between the two very *likely*. Still, "homology modelling" is the standard term used in this context although "comparative modelling" may be used equivalently and is the more accurate term [8, 66, 67]. Protein structure is more strongly conserved in evolution than sequence [50] so even a moderate level of similarity over the entire sequence suffices to be confident of high structural similarity – although there are a few notable examples where this does not hold, see e.g. Ref. [51] where two proteins are engineered at 88% sequence identity but with completely different (α as opposed to α/β) folds or Ref. [52] for two naturally occurring proteins of 40% sequence identity and different folds. The general rule for natural proteins though is that sequence similarity means high structural similarity and one of the main challenges is to properly incorporate information from remote homologues [8].

If no such structures with sequences of sufficiently high similarity to the query sequence are detected, structure has to be predicted *ab initio*. This does not mean that no structural information from other proteins enters the prediction but that information is usually only local and also less reliable. This branch of protein structure prediction is therefore more challenging than template-based homology modelling [68, 69].

The bi-annual Critical Assessment of Techniques for Protein Structure Prediction (CASP) has witnessed remarkable progress in both these categories over the last 15 years [70–74]. For this assessment experimentalists agree to hold back recently resolved structures and theorists are invited to send in their predictions. One method that repeatedly performed very well in CASP is Rosetta [10, 72, 75]. There, starting from a sequence, the first step is to predict secondary structure and create a library of structure fragments for that particular sequence based on sequence and secondary structure similarity. These fragments are then assembled into complete, folded protein structures and inclusion of different fragments or local movements are proposed according to a Monte Carlo (MC) scheme.

The fragment assembly step results in a set of very many coarse-grained protein structure guesses which is expected to contain a few candidates that are close to the native structure. If computation time were not an issue all these candidates could be fine-grained, i.e. omitted side chains included [56], and optimised again. However, only those guesses that are already close enough to the correct structure will profit from this refinement, the vast rest will remain trapped in their respective (wrong) folds. It is hence wasteful of computer resources to treat all candidates to further optimisation. In this first part of the thesis, I therefore investigate the task to identify the few good candidates contained in the large coarse-grained sets [76].

This chapter is organised as follows: After a short presentation of the research context and background material, Section 2.2 introduces the benchmark proteins used in this study (Subsection 2.2.1) and summarises the methods of profile prediction (Subsection 2.2.2). Subsections 2.2.3 and 2.2.4 present the filtering results when using Rosetta to predict protein structures, Subsection 2.2.5 the results for structure predictions downloaded from the CASP8 website. The chapter ends with a discussion (Section 2.3).

2.2 Structure Selection

Simulation of detailed side-chains in realistic potentials faces two serious drawbacks: For one thing it is a very computation-intensive task that requires immense computing resources. The other, more fundamental, problem is that today's all atom potentials or force fields are optimised for fully folded native proteins and thus can only faithfully model the vicinity of these structures [23]. The vast space of unfolded or only partially folded structures can thus not be expected to be represented as accurately.

This makes such detailed potentials unsuitable for following the entire folding process – and even more so for predictive folding where no parameters can be tuned in favour of the native structure. Successful structure predictors therefore ignore the folding process and concentrate on the final structure. For this reason, the time series of protein conformations encountered in, for example Rosetta, cannot be considered as a folding trajectory. Instead, entire fragments of the protein structure are replaced in a single step which speeds up the sampling of conformation space and increases the chances to hit on a structure that has at least the overall correct fold.

Thousands of candidate structures are produced in the coarse-grained step, which only describes the protein's backbone and some interactions such as steric repulsion. It is necessary to produce very many structures at this stage as the coarse-grained potential may not produce close guesses on every occasion. For single structure predictions that were entered into CASP, the Rosetta group indeed fine-grained all these candidate structures [72], but this is not feasible for high-throughput predictions. So, as has been mentioned before, the task is to select only promising candidate structures for high-resolution refinement [77] in phenomenological [78] or physics-based [79] force fields.

The selection step may involve ranking candidate structures by the energy function used to produce the low-resolution structures. Therefore an option to improve the selection is to define better energy functions for scoring of low-resolution candidate structures [9]. Another approach, which appears to be promising, is to perform a clustering of the structures by their pairwise root mean square distances (RMSDs) and then consider the largest clusters [10] or the clusters of lowest energy [11]. This is based on the notion that while the coarse-grained model may not succeed in discriminating the single best structure by energy, it will still on average create many conformations in the vicinity of the native structure which can be detected by clustering using pairwise similarities. Clustering by distance matrices and identifying the cluster of lowest energy has also proved successful in the reconstruction of protein structures from highly approximate backbone torsion angles [80].

Related to the clustering approach is the definition of a meta-scoring function based on the correlation of scoring functions that are weakly funneled towards the native state [81]. This has been applied to the ranking of predicted protein models [82] which is similar in spirit to the selection of candidate structures. The meta-method of detecting similarities and cleverly combinining predictions of different methods has also been successful in recent rounds of CASP [83, 84]. If sparse experimental data is known, such as NMR chemical shifts, their inclusion as constraints on protein structure substantially improves prediction [85, 86].

Another promising means to select structure candidates for refinement is the use of structural profiles such as the effective connectivity, Eq. (1.4). As will be discussed in more detail below, filtering by a predicted profile outperformed filtering by Rosetta's low-resolution energy and clustering by RMSDs in most cases [76], irrespective of whether candidate structures were produced using the Rosetta suite or downloaded from the CASP8 (8th Critical Assessment of Techniques for Protein Structure Prediction) server [87] and thus came from various prediction methods. Structural profiles can be determined from known structures and used to efficiently compare them [63] but, most importantly in the context of this chapter, the structural profile of a protein's native state can also be predicted to good accuracy from its amino acid sequence. Prediction of one-dimensional structural profiles is above all much easier than the prediction of residue-residue contacts, i.e. two-dimensional contact maps. The predicted profile can then be used as a target which is compared to every single profile computed from the candidate structures. Only those candidate structures with profiles that are similar to the predicted target are selected as suitable for refinement. I compare filtering by either exact or predicted profiles to filterings obtained from established methods and thereby show that the predicted version is already comparable to, if not better than, the methods of clustering or selection by energy but further improvement can be expected from improvements in profile prediction.

2.2.1 Benchmark Structures

The different filtering methods were tested on two different sets of proteins. For the first protein set (listed in Table 2.1) structure sets containing 10,000 candidates for each target were produced using the standard Rosetta *ab initio* protocol [10]. In order to assess the filtering performance of the different methods, sequences with known structures were used. Thus, for a realistic prediction scenario, these structures as well as close homologues (judged by sequence similarity) were excluded for the generation of fragments.

The lengths of protein structures in this set range from 46 to 209 amino acids. For some proteins (1shg, 1r69, 1p9yA and 1gk9) the lengths of structures in PDB files differed from sequence lengths in the FASTA files obtained from the PDB website [35] as experiments occasionally fail to properly resolve the rather unstructured tail regions. In those cases, sequences were shortened for fragment prediction and assembly to the corresponding structure lengths. Decoy generations took 1-3 days per protein, depending on sequence length, on a modern desktop computer.

The second set of benchmark structures consists of models submitted to CASP8 [87]. Starting from the server-only *ab initio* predictions those targets for which experimentally resolved structures were incomplete or ruptured were discarded to simplify comparison to predictions. Only missing end termini were allowed. The list of suitable target proteins (29 of 69) are given in Table 2.2. These proteins range from lengths of 69 to 533 amino acids. They are therefore somewhat longer than the proteins that could reasonably be studied by creating decoy sets myself. Furthermore, these models came from several different groups using different generation protocols and the results are in that respect more general. For these CASP targets only 100 to 300 models per target were available (instead of the decoy sets of 10,000) but these were the models selected, by the respective groups, for submission.

PDBid	description	length	class
1pv0	Sda antikinase	46	α
1gb1	immunoglobulin binding domain of protein G	56	$\alpha + \beta$
1shg	SH3 domain	57	β
1jic	sso7d protein	62	β
1r69	N-terminal domain of phage 434 repressor	63	α
1c9oA	chain A of cold shock protein	66	β
1mjc	major cold shock protein	69	β
1fgp	minor coat protein G3P	70	β
1ubq	ubiquitin	76	$\alpha + \beta$
1oqp	C-terminal domain of caltractin	77	α
1btb	barstar	89	α/β
1p9yA	single chain of trigger factor	117	$\alpha + \beta$
2imf	2-hydroxychromene-2-carboxylate isomerase	203	α
1volA	chain A of transcription factor IIB	204	α
1ix9A	single chain of manganase(III) superoxide dismutase mutant Y174F	205	α
1f5x	autoinhibited Dbl homology domain	208	α
1gk9A	chain A of penicillin acylase enzyme-substrate complex	208	α
1by1	Dbl homology domain from beta-PIX	209	α

Table 2.1: Protein target structures used to test selection performance of various filters, sorted by length. Protein descriptions are taken from the Protein Data Bank (PDB) website [35], class refers to Structural Classification of Proteins (SCOP) [47]. If the PDB file contained more than one chain, the chain identifier is given, such as in 1c9oA. Candidates were predicted using the Rosetta suite [10].

PDBid	description	length
3dex	SAV_2001 protein	69
3ded	C-terminal domain of probable hemolysin	77
3d7i	hypothetical protein MJ0742	83
2vsv	PDZ domain of human rhophilin-2	87
3dm3	domain of a replication factor A protein	88
3dm4	primosomal replication protein	88
3df8	possible HxlR family transcriptional factor	99
2k54	protein Atu0742	112
3dkz	Q7W9W5_BORPA protein	115
3dai	bromodomain of human ATAD2	120
3d0j	CA_C3497, unknown function	128
3dn7	cyclic nucleotide binding regulatory protein	136
3dmb	putative general stress protein 26	136
3di5	DinB-like protein (NP_980948.1)	140
3d8p	acetyltransferase of GNAT family	140
3cyn	human GPX8	164
3d3o	effector domain of the putative transcriptional regulator IclR	166
3d5n	Q97W15_SULSO protein	168
3d7l	protein lin1944	192
3dlm	Tudor domain of human Histone-lysine N-methyltransferase	198
3dc7	protein Q88SR8	203
3dlc	putative S-adenosyl-L-methionine-dependent methyltransferase	205
2vx2	human enoyl coenzyme	246
3da2	human carbonic anhydrase 13 in complex with inhibitor	252
3d19	conserved metalloprotein	254
3dao	putative phosphatase	255
3dsm	surface layer protein BACUNI_02894	317
2vuw	kinase domain of human haspin	319
3do6	putative formyltetrahydrofolate synthetase	533

Table 2.2: Protein target structures used to test selection performance of various filters, sorted by length. Protein descriptions are taken from the Protein Data Bank (PDB) website. Candidates were models submitted to CASP8.

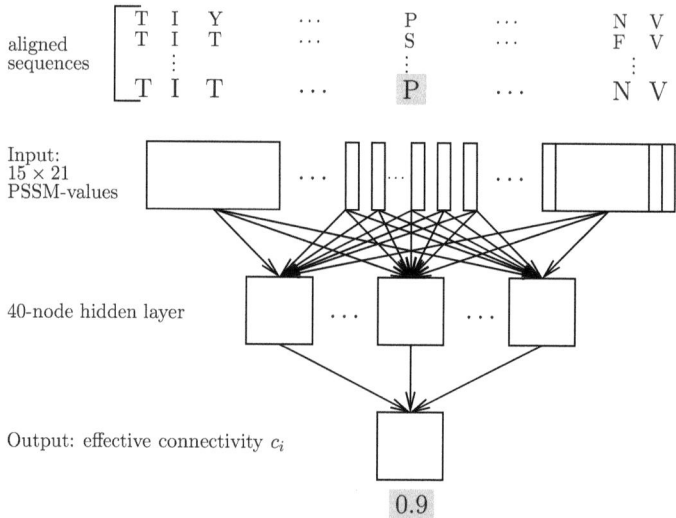

Figure 2.1: Artificial Neural Network (ANN) to predict structural profile from sequence.

2.2.2 Prediction of Structural Profiles

As a proof of principle I first used exact profiles for filtering, i.e. effective connectivities (ECs) computed from the target structure using Eq. (1.4). Obviously, this definition requires knowledge of the native structure and is thus of no use in *ab initio* structure prediction. But, as mentioned in Section 1.3, there is a good correlation between structural profiles and sequence hydrophobicities [7]. This correlation can now be exploited to predict the profile from the amino acid sequence with good accuracy by using feed-forward artifical neural networks (ANN).

The correlation between the structure's profile and *optimal* sequence hydrophobicity, however, is still greater than that between the profile and an individual sequence's hydrophobicity [7]. Predictions can therefore be improved if the optimal sequence is approximated by an evolutionary average obtained from a sequence alignment of the sequence in question. However, instead of attempting to directly determine the optimal sequence or a single evolutionary average, a PSI-BLAST [88] search is performed to yield position-specific scoring matrices (PSSMs) for each sequence position. These PSSMs encode the probability (in log-odds) of each amino acid species to occur at a given position which are then used to predict the structural profile. This approach has first been developed in the context of secondary structure prediction [89] and was adapted to structural profiles in the diploma thesis of Jonas Minning [90].

The ANN then uses as input the PSSMs for a sequence window of 15 amino acids centered at the position for which the structural profile entry is to be predicted (see Figure 2.1). To predict the entire profile this window slides over the sequence entering amino acid scores into the ANN or, if part of the window extends the sequence's end, the information that there is no amino acid. There are therefore 15 (i.e. size of window) times 21 (i.e. number of canonical amino acids plus empty space) input nodes to the ANN, each giving the scoring of every amino acid species at that position in the sequence. The ANN further consists of a hidden layer of 40 neurons and a single output neuron. The activation function of the hidden layer is the hyperbolic tangent and the output is linear. The output neuron then returns a single real value, the predicted profile entry for the central amino acid of the window.

The ANN was trained on a representative subset of the PDB of 300,000 residues in total to minimise the squared differences between exact EC and prediction with early stopping to avoid overfitting. No difference in prediction quality could be observed depending on the inclusion or omission of sequence homologues, so the possibility of a hidden homology modeling is ruled out.

2.2.3 Distribution of Rosetta Candidate Structures

Quality of candidate structures can be assessed by different measures. Here, two different distance measures are tested, C_α-RMSD and TM-score (also based on C_α atoms) [91]. The root mean square deviation (RMSD) after a rotation which optimally superimposes the two structures [92–94] is perhaps the most intuitive measure of protein similarity and adequate for closely related structures but carries less information for dissimilar structures [95]. The TM-score does not suffer from this disadvantage and can detect even quite weak similarities between structures. Its definition is less intuitive,

$$\text{TM-score} = \max\left(\frac{1}{L}\sum_{i=1}^{L}\frac{1}{1+\left(\frac{d_i}{d_0}\right)^2}\right), \quad (2.1)$$

with L being the length of both structures (there is also a more general version for structures of unequal size). The distance between C_α-atoms at position i in both structures is denoted as d_i, while d_0 is a heuristically determined parameter, $d_0 = 1.24\sqrt[3]{L-15} - 1.8$. The maximisation is again over all possible rotational superpositions of the structures. While RMSD is minimal for highly similar structures and dependent on length, TM-score lies between 0 and 1, is length-independent and maximal for the most similar structures. Furthermore, TM-score puts a larger weight on those parts that are similar so weaker similarities can be detected. However, the results for filtering performances and ranking of methods were essentially equivalent when using either RMSD or TM-score. The more intuitive RMSD was therefore used to rank filtering methods in the first part and TM-score was computed only to ensure that both measures by and large agreed. In the second part, in which the protein set from CASP8 containing longer structures was investigated, the length-independent TM-score was used.

A second question is how to compare distributions (of either RMSD or TM-score) of structure sets or candidate selections. The entire distribution carries most information but is also difficult to compare quantitatively. The average RMSD (or average TM-score) to the native structure is not a very suitable quantity as the main goal is to find the few structures of very low RMSD (or high TM-scores) even if some bad structures are contained in the selection.

For the initial structure sets created by Rosetta I therefore show full distributions of RMSD and TM-score. Two typical cases are given in Fig. 2.2, the majority of plots can be found in the Appendix to this thesis. As larger TM-scores stand for more similar structures, all figures in this thesis report 1 − TM-score, so "good" structures appear towards the left of the plot and "bad" structures towards the right to facilitate comparison to RMSD values.

For structure selections by different filtering methods I show distributions only for example proteins which are either typical cases or noteworthy exceptions. Instead, the structure of lowest RMSD value in the selection and the number of structures below a certain RMSD threshold will be listed. Experience shows that a threshold for a good subsequent refinement should be at an RMSD of about 3-4 Å (or a TM-score threshold of 0.6) [77, 91]. However, distributions differ for different proteins and for some the decoy set does not contain any such good structures. A protein-dependent threshold therefore becomes necessary and all structures with an RMSD one standard deviation lower than the mean RMSD ($z_{\text{RMSD}} < -1$, or mean TM-score, $z_{1-\text{TM-score}} < -1$) are deemed good.

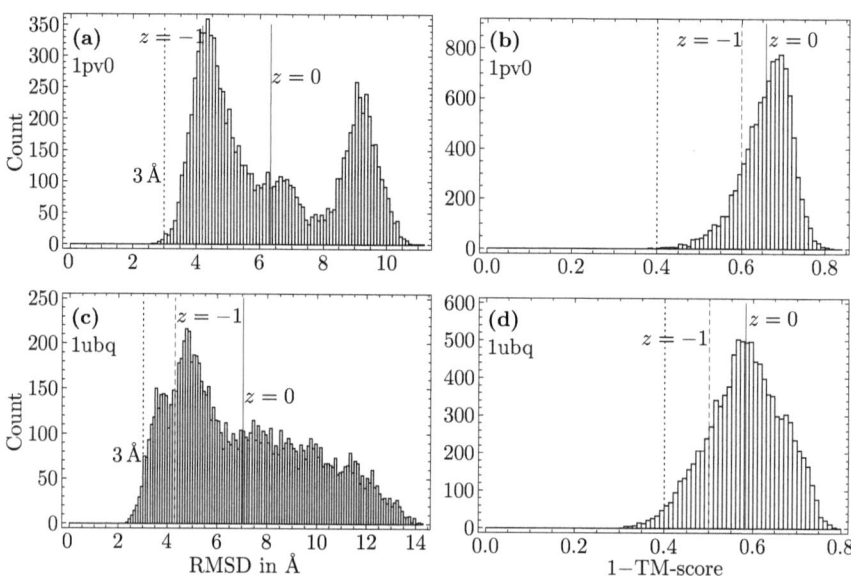

Figure 2.2: Distribution of RMSD and 1−TM-score for Sda antikinase (PDB id. 1pv0, length 46) and ubiquitin (PDB id. 1ubq, length 76).

For most of the smaller proteins, i.e. those with less than 200 amino acids, RMSD values of candidate structures to target structures were found to lie between 2.5 Å and 14 Å. TM-score typically ranged from 0.2 to 0.8 (see Fig. 2.2 and Appendix). The dotted lines correspond to an RMSD threshold of 3 Å and a TM-score threshold of 0.6 (1−TM-score= 0.4). All candidates to the left of these lines are very good and likely to converge even further in all-atom refinement. The red line marks the mean of the distribution and the dashed red line one standard deviation below the mean ($z = -1$) as a measure of relatively good structures within a given decoy set.

Notable exceptions in decoy distributions are found for 1p9yA and 1r69 (see Appendix). The former distribution, 1p9yA, contains some structures of very high RMSD (ca. 25 Å) whereas for the latter, 1r69, the entire RMSD distribution is fairly good and TM-scores even reach values of 0.9. For 1r69 even a random selection of structures might be suitable for refinement to high-resolution structures. The reason for these abnormal cases appear to be the complex structure of 1p9yA on the one hand, where two distant β-hairpins have to meet, and the fact that 1r69 on the other hand had been used for calibration of the Rosetta score [96], explaining the very good decoy distribution in this case.

For larger proteins with 200 amino acids and more the decoy sets deteriorated drastically and RMSD distributions ranged from 7 Å to 30 Å, some targets, 1ix9 and 1gk9, did not even contain any candidates below an RMSD of 10 Å. For TM-score distributions this meant values between 0.1 and 0.5, for some proteins only up to 0.4. These candidates are practically unrelated to the target structure and no meaningful comparison of filtering methods was possible. Even the more sensitive TM-score was unable to detect weak signals of similarities between those conformations and the target structure.

It is also worth noting that while "good" RMSD distributions for small proteins showed some structure and more than one minimum, RMSD distributions for longer proteins typically become Gaussian in shape. The reason for this behaviour may be the very same effect that is exploited for clustering: For small proteins some regions of conformation space with typical structures are sampled more frequently, and if equilibrium is assumed these distribution maxima would correspond to free energy minima. For longer proteins conformation space becomes so large that mostly unrelated structures are sampled that follow a Gaussian distribution around some typical length-dependent RMSD.

For each query sequence 10,000 structures were generated. For the longer structures the number of candidates necessary to expect one structure with RMSD ≤ 5 Å were estimated by approximating distributions as Gaussian with mean and variance calculated from the structure sets. This resulted in numbers up to 10^{12} structures for e.g. 1ix9 (see Appendix). The large number of structures that would have been required was one of the reasons to instead turn to CASP8 models where good predictions of longer proteins could be found.

2.2.4 Filtering of Rosetta Candidates

Three different methods were considered for narrowing down the number of structures from the coarse-grained set, one of which was filtering by, either exact or predicted, effective connectivities. Another way is to use the same score as was used in the candidate structure generation, i.e. the standard low-resolution score of Rosetta [10],

$$E_{\text{score}} = E_{\text{env}} + E_{\text{pair}} + E_{\text{vdw}} + E_{\text{hs}} + E_{\text{ss}} + E_{\text{sheet}} + E_{\text{r-sigma}}. \qquad (2.2)$$

These energy terms all operate on the coarse-grained backbone structure and stand for residue interactions with the environment (i.e. solvation energy E_{env}), pairwise interactions of residues, E_{pair}, and van der Waals-interaction in the form of steric repulsion, E_{vdw}. The last four terms, E_{hs}, E_{ss}, E_{sheet} and $E_{\text{r-sigma}}$ all account for stacking and packing of secondary structure elements.

The filtering score based on the effective connectivity, see Eq. (1.4), reads as

$$\Delta_{\text{EC}}(j) = \sum_{i=1}^{L} \left| t_i - c_i^{(j)} \right|^\alpha \qquad (2.3)$$

where the index i runs over all L amino acids and index j enumerates the candidate structures of the coarse grained set. The target EC **t** with entries t_i can be either the predicted or the exact profile and $\mathbf{c}^{(j)}$ denotes the EC computed from candidate structure j. The exponent α is set to 2 for the following results but varying it between 0.5 and 4 made hardly any difference. Applying these filters to the structure set the $x\,\%$ structures of lowest score were selected, with $x \leq 2$.

Another less obvious parameter that enters the filtering by EC is the contact threshold r_c used for determining the contact map in Eq. (1.1). The choice of parameter enters into both the computation of exact structural profiles and, more indirectly, into the training of the ANN for prediction. Both prediction quality of ECs and filtering quality depend on this but the effect varied for different proteins. For the results reported here an intermediate value of $r_c = 8.5\,\text{Å}$ has been used as contact threshold for the distances between C_α-atoms.

The last method used for selection of structures was clustering by pairwise RMSDs between candidates. The idea behind clustering is that it can find representative structures of similar configurations and thus identify highly populated free energy minima. Assuming that, while the absolute value of the coarse-grained energy may not be very accurate and the depth of the native basin not well-represented, the width still may be preserved, configurations would be more densely sampled around the native basin and large clusters of low pairwise RMSDs would contain near-native structures. In order to compare this method to filtering by the above scores, the clustering procedure was tuned such as to return a largest cluster of 200 structures, corresponding to 2 % of the total set of 10, 000. The method used here is a very simple clustering algorithm and involves five steps:

(1) Compute pairwise RMSDs between all candidate structures.

(2) Choose an RMSD threshold.

(3) Find the configuration with most neighbours, i.e. most configurations within the RMSD threshold. This is the centre of the largest cluster.

(4) Remove the largest cluster (the centre and all its neighbours).

(5) Continue at step (3) with the remaining configurations.

This procedure is repeated to extract the ten largest clusters. The RMSD threshold of step (2) is determined in a binary search to obtain a largest cluster of approximately 200 structures.

The RMSD distribution of the largest cluster and the RMSD distribution of the cluster with lowest average Rosetta energy (Eq. 2.2) are compared to those obtained by filtering by either of the scoring functions. Additionally, the centres of all ten largest clusters are compared to the native structure.

A first test of the filtering based on scoring functions is the correlation of scores to RMSD and their funnelling towards the native structure. Scatter plots for the Rosetta score (2.2) and the EC-score (2.3), for exact Δ_{EC} and predicted Δ_{predEC} profiles, over RMSD are shown in Fig. 2.3 for two different proteins, ubiquitin (PDB id. 1ubq) and the SH3 domain (PDB id. 1shg).

A configuration's Rosetta score (low-resolution energy function) is usually correlated to its RMSD from the native structure (see Fig. 2.3 **(a)**) but funneling towards the native structure is not optimal as some structures of RMSDs between 4 Å to 5 Å have lowest energies although structures of RMSD down to 2 Å exist in the coarse-grained structure set. Since only the structures of best energies will be selected from the set, the funnelling property is even more impor-

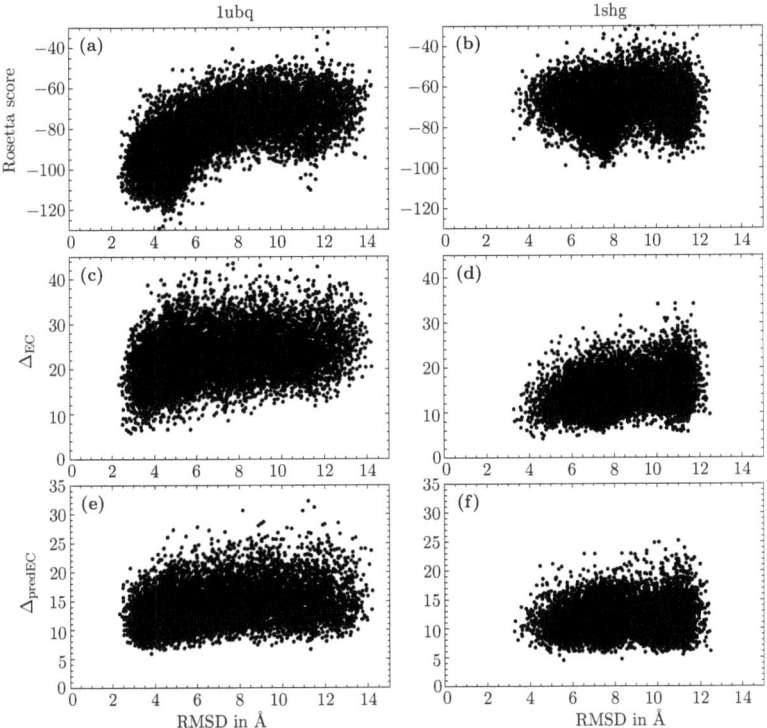

Figure 2.3: Correlation of the Rosetta score and the EC-score for exact profile, Δ_{EC}, and predicted profile, Δ_{predEC}, to RMSD for ubiquitin (PDB id. 1ubq) and the SH3 domain (PDB id. 1shg).

tant than good correlation reaching up to structures of high RMSDs. In some cases (Fig. 2.3 **(b)**) the Rosetta score fails more dramatically and is very misleading.

The scoring function Δ_{EC} is more reliable and always places structures of very low RMSD among the top scoring candidates. For ubiquitin, there is a clear funnel when using the exact profile (Fig. 2.3 **(c)**), for the SH3 domain this property is weaker (Fig. 2.3 **(d)**) but still considerably better than for the Rosetta score. Scatter plots for scoring using the predicted structural profile show less funnelling than those for the exact profile. The overall correlation may even be weaker than for the Rosetta score but better for structures of low RMSD (Fig. 2.3 **(e)**) or at least do not lead towards false minima (Fig. 2.3 **(f)**).

A more stringent assessment of filtering quality than the scatter plots of Fig. 2.3 is to look at the top structures actually selected by the different methods. In Fig. 2.4 the distribution of **(a), (c)** RMSD and **(b), (d)** 1−TM-score is given for two example proteins, namely chain A of the cold shock protein (PDB id. 1c9oA) and ubiquitin (PDB id. 1ubq). A filtering method has

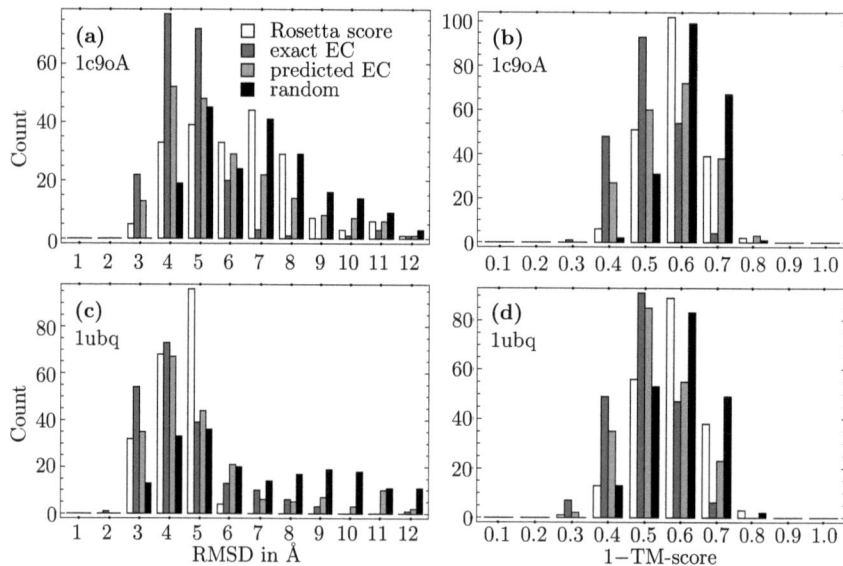

Figure 2.4: RMSD and 1−TM-score distribution of filtered structures for Rosetta score, EC-score with exact or predicted profiles and random selection for chain A of the cold shock protein (PDB id. 1c9oA) and for ubiquitin (PDB id. 1ubq).

performed the better the more structures of low RMSD (or low values for 1−TM-score) are contained in the set. In both cases filtering by exact EC clearly beats filtering by the Rosetta score and filtering by predicted EC is comparable or slightly better when looking at the RMSD distributions and quite clearly better when judging by TM-scores. For better comparison a histogram for randomly selected structures has been included as well, which, as expected, contains fewer good structures than either of the two filtering methods.

The results for all other proteins (see Table 2.1) are summarised in Table 2.3 and are based on RMSD distribution after filtering. Those proteins with lengths above 200 amino acids are omitted because, as has been argued above, the starting distributions are too bad to extract any meaningful results. The second to sixth column in Table 2.3 show the number of good structures and, in brackets, the RMSD value of the best structure in the selection by exact EC, predicted EC, the Rosetta score, the largest cluster C_1 and the cluster of lowest average Rosetta score C^*. Structures are defined to be (relatively) good if they lie at least one standard deviation below the mean of the entire RMSD distribution and the number of good structures is denoted by $N(z_{\text{RMSD}} \leq -1)$. Usually, the method that contains most "good" structures also returns the lowest-RMSD structure. The last column of Table 2.3 gives the lowest RMSD of the entire structure set for the respective protein for comparison.

Filtering by exact EC in most (9 out of 12) cases outperforms filtering by the Rosetta score and for two of the proteins where this is not so (PDB ids. 1gb1 and 1mjc) the results are very close

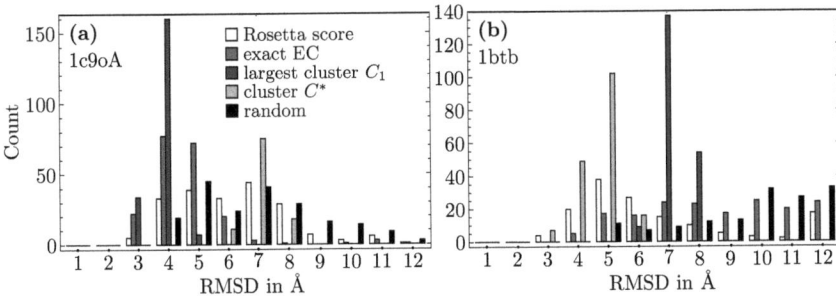

Figure 2.5: RMSD distribution of filtered structures including filtering by clustering for the two proteins, chain A of the cold shock protein (PDB id. 1c9oA) and barstar (PDB id. 1btb) for which clustering was successful.

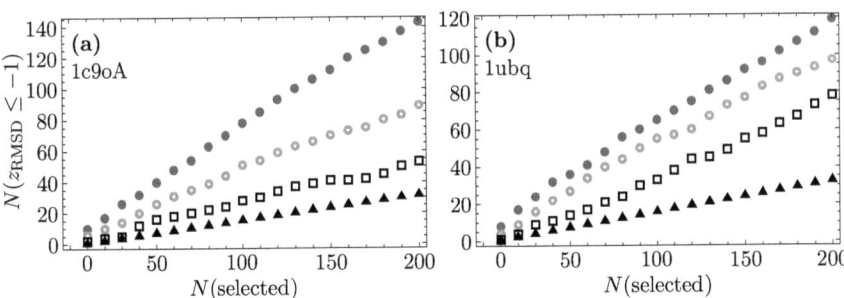

Figure 2.6: Number of good structures in dependance of number of selected structures for chain A of the cold shock protein (PDB id. 1c9oA) and for ubiquitin (PDB id. 1ubq). Red circles denote selection by exact EC, orange open circles by predicted EC, white squares by the Rosetta score and black triangles a random selection of structures.

and which method returns more good structures depends on how many candidates are selected from the set. Predicted EC is still somewhat better than the Rosetta score, it returns better structures six times versus five. There is one tie (PDB id. 1r69) where one method returns more of the "good" structures and the other the best single structure. The fifth and sixth column give the results for clustering. These usually contain only few of the good structures with two notable exceptions being the largest cluster for the cold shock protein (PDB id. 1c9o) and the cluster of lowest Rosetta energy for barstar (PDB id. 1btb). Histograms for these two proteins are given in Fig. 2.5. It stands out that for barstar (Fig. 2.5 **(b)**), for which the cluster of lowest energy is exceptionally good, the Rosetta energy on its own also gives very good results so clustering only serves to concentrate these good structures. This, however, is also the disadvantage of clustering, as it narrows down the selection quite severely and if it errs, frequently contains no good structures at all.

PDBid	EC	predEC	Rosetta score	C_1	C^*	minRMSD
1pv0	72(2.9)	26(3.1)	10(3.5)	0(6.1, 199)	26(3.7, 58)	2.6
1gb1	151(1.6)	9(1.7)	186(1.5)	0(3.8, 196)	91(1.5, 91)	1.5
1shg	58(3.6)	38(4.0)	7(4.9)	0(6.9, 209)	0(7.1, 49)	3.3
1jic	50(3.8)	23(5.8)	4(6.4)	0(10.9, 200)	0(10.7, 164)	3.1
1r69	37(1.6)	15(2.2)	34(2.3)	0(3.0, 191)	1(3.0, 70)	1.6
1c9oA	143(2.9)	89(2.9)	53(3.2)	201(2.8, 201)	0(5.6, 104)	2.8
1mjc	127(2.8)	80(2.8)	138(2.9)	166(4.3, 201)	166(4.3, 201)	2.8
1fgp	102(5.9)	25(9.4)	28(8.7)	0(10.7, 196)	21(9.8, 185)	5.9
1ubq	119(2.4)	97(2.7)	78(2.7)	26(3.5, 196)	0(4.6, 42)	2.3
1oqp	76(4.1)	29(4.2)	51(4.2)	84(4.5, 200)	0(5.4, 70)	3.7
1btb	70(3.7)	87(3.4)	109(3.1)	186(6.2, 200)	174(3.1, 174)	3.1
1p9yA	40(5.7)	30(6.1)	17(5.4)	27(8.1, 198)	25(6.2, 198)	4.7

Table 2.3: Quality of different filtering methods. Number of structures with RMSD values smaller than one standard deviation below the mean $N(z_{\text{RMSD}} \leq -1)$ for various selection methods and minimum RMSD in Å (in brackets). For the clustering method, the second number in brackets is the cluster size. The largest cluster is denoted as C_1, the cluster of lowest average Rosetta score as C^*. The last column gives the overall minimum RMSD in Å. For 1mjc C_1 and C^* coincide, for 1p9yA there are two equally large clusters C_1 one of which is also C^*. In every line the best method, disregarding the case of the exact EC, is underlined, which is usually the method achieving highest $N(z_{\text{RMSD}} \leq -1)$. For 1mjc and 1oqp, the Rosetta score's results are considered as better than results from C_1 although $N(z_{\text{RMSD}} \leq -1)$ was higher for C_1 because the overall RMSD distributions are better. Similarly, for 1btb C^* contains fewer "good" structures (as measured by $N(z_{\text{RMSD}} \leq -1)$) than C_1, however, the RMSD distribution is better in that it contains more very good structures. For 1r69 the case is left undecided between predicted EC and Rosetta score. The proteins are sorted by length.

As has been mentioned before, the (relative) number of good structures contained in the selection also depended on the number of structures drawn from the entire set. Figure 2.6 shows the quality of selections with increasing size for two example proteins and also compares them to the number of good structures expected in a random sample. The latter has been estimated as $N(\text{selected})|G|/|A|$, where G is the set of all good structures and A the entire decoy set. The number of selected structures, $N(\text{selected})$ was always much less than the number of good structures, $|G|$, in the entire set. It holds $N(z_{\text{RMSD}} \leq -1) = |G \cap S_f|$ with S_f the set of structures selected by filter f (exact EC, predicted EC or Rosetta score). For the proteins shown here the numbers of good structures in the sets selected by exact or predicted EC are always higher than those chosen using the Rosetta energy. However, it appears that the EC curves flatten somewhat with increasing selection size while the Rosetta score catches up. This is important if, due to limited computer resources, only a smaller number of structures is to be passed on to refinement. Similarly, filtering by (exact or predicted) EC is often more effective at identifying a structure of very low RMSD from the structure set (see Table 2.3, numbers in brackets).

PDBid	ρ_c	predEC	mEC, $\rho_c = 0.8$	mEC, $\rho_c = 0.9$	exact EC
1pv0	0.64	26(3.1)	25(3.1)	34(3.1)	72(2.9)
1gb1	0.30	9(1.7)	67(1.7)	102(1.7)	151(1.6)
1shg	0.62	38(4.0)	41(4.0)	43(4.0)	58(3.6)
1jic	0.72	23(5.8)	22(6.2)	27(6.2)	50(3.8)
1r69	0.54	15(2.2)	21(2.2)	25(1.6)	37(1.6)
1c9oA	0.80	89(2.9)	90(2.9)	116(2.9)	143(2.9)
1mjc	0.86	80(2.8)	-	89(2.8)	127(2.8)
1fgp	0.35	25(9.4)	37(8.5)	49(8.0)	102(5.9)
1ubq	0.30	97(2.7)	122(2.5)	126(2.5)	119(2.4)
1oqp	0.69	29(4.2)	32(4.1)	39(4.1)	76(4.1)
1btb	0.48	87(3.4)	87(3.4)	82(3.2)	70(3.7)
1p9yA	0.69	30(6.1)	30(6.1)	33(5.7)	40(5.7)

Table 2.4: Filtering by interpolated structural profiles. (Column 2) Pearson's correlation coefficient ρ_c between predicted and exact EC, (Columns 4 and 5) Interpolated EC (mEC) with fixed correlation to exact EC (0.8 resp. 0.9). Data reported are number of good structures $N(z_{\text{RMSD}} \leq -1)$ and lowest RMSD in the selection (in brackets). (Columns 3 and 6) Data repeated from Table 2.3 for predicted and exact EC for comparison. The proteins are sorted by length.

As filtering by exact EC is almost always more effective than filtering by the predicted structural profile, with the single exception being barstar (PDB id. 1btb), it is interesting to compare the quality of profile prediction to filtering performance. Table 2.4 therefore gives Pearson's correlation coefficients between exact and predicted profiles,

$$\rho_c = \frac{\sum_{i=1}^{L} (t_i - \langle t \rangle)(p_i - \langle p \rangle)}{\sqrt{\sum_{i=1}^{L} (t_i - \langle t \rangle)^2 \sum_{i=1}^{L} (p_i - \langle p \rangle)^2}}. \quad (2.4)$$

Here, t_i stands for the vector entries of the exact profile with mean $\langle t \rangle$ and p_i for those of the predicted profile with mean $\langle p \rangle$. Incidentally, both profiles are normalised such that their means are equal to 1. The correlation values for the smaller proteins are given in the second column of Table 2.4 ranging from $\rho_c = 0.3$ to $\rho_c = 0.86$. While these values vary considerably, it is important to stress that they do not systematically deteriorate with length. Values for the large (more than 200 amino acids) proteins were also similar and lay between $\rho_c = 0.58$ and $\rho_c = 0.78$. The bottleneck for larger structures thus is not prediction of profiles but indeed generation of candidate structures. Table 2.4 also shows that the dependance of filtering on prediction quality is not the same for different proteins. Domain B1 of protein G (PDB id. 1gb1), for which filtering by predicted EC exhibited the worst performance when compared to filtering by exact EC, also has one of the lowest correlation coefficients – but so does ubiquitin (PDB id. 1ubq) for which filtering was quite good.

Higher correlation values were simulated by creating "mixed ECs", i.e. linear interpolations between exact and predicted profile, to yield profiles of a prescribed correlation (either $\rho_c = 0.8$ or $\rho_c = 0.9$). Results for filtering by these interpolated profiles are given in the fourth and fifth column of Table 2.4. For ease of comparison numbers for the predicted and exact profiles are repeated in the third and sixth column. For almost all proteins filtering performance increases

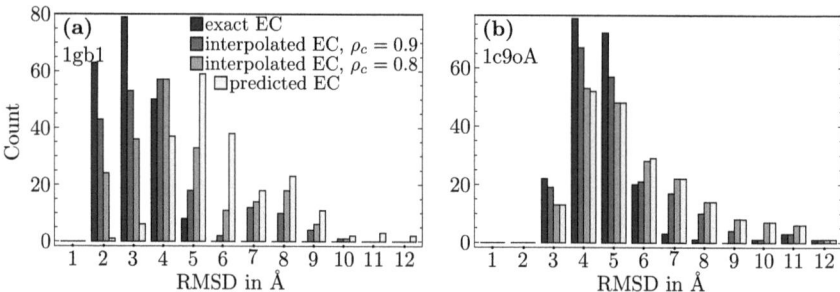

Figure 2.7: Distribution of filtered structures for interpolated ECs for **(a)** the B1 domain of protein G (PDB id. 1gb1) and for **(b)** chain A of the cold shock protein (PDB id. 1c9oA).

for increasing correlation of interpolated profiles, the only exception being barstar (PDB id. 1btb) for which the predicted profile gave better results than the exact one. RMSD distributions for two more typical proteins when filtered by interpolated profiles are given in Fig. 2.7. If, for example, prediction quality of domain B1 of protein G (PDB id. 1gb1) is artificially increased to a value of $\rho_c = 0.8$, filtering performance is comparable to the other proteins (see Table 2.4 and Fig. 2.7 (a)). Filtering by profiles that had good correlations to begin with are usually also improved by increasing correlation (see Fig. 2.7). But still, even for fixed correlation and even when set into relation with numbers for exact ECs, filtering performance varies considerably. Quality of profile prediction can therefore not be the only determinant of quality of filtering, protein-specific features or the structure distribution in the entire candidate set also play an important role.

In particular, filtering by structural profiles compares favourably to filtering by the Rosetta score for decoy sets of lower quality (see Table 2.3) where still relatively good structures can be extracted. This is important as candidate sets deteriorate with increasing sequence lengths whereas prediction of structural profiles does not systematically do so. If, however, candidate sets are of very poor quality, as is the case for the rather small sets for proteins of lengths greater than 200 amino acids, none of the filtering methods investigated is capable of extracting any meaningful subset of structures. Filtering by predicted profiles thus has the potential to improve selections for more difficult structures (such as chain A of the trigger factor 1p9yA with its complex β-topology) but meets its limits for too long proteins.

It has been mentioned above that clustering and picking a single cluster has the disadvantage of narrowing down the structure set such that the selected cluster frequently contains not a single good structure. Using only the largest cluster or only the cluster of lowest average energy is therefore quite risky although it may give very good results in some rare cases (see Fig. 2.5). Changing the clustering RMSD threshold such that the largest cluster for instance contained 500 instead of 200 structures brought no significant changes.

The opposite approach is to aim at covering the relevant parts of conformation space and only consider representative structures in the form of the centres of all ten of the largest clusters. This was investigated as well and compared to the ten top-ranked structures by exact or predicted EC. Results of this test, i.e. numbers of good structures and minimum RMSD among the ten selected, are given in Table 2.5. The exact EC returns the largest number of good structures for

PDBid	$C_{centres}$	EC	predEC
1pv0	4(3.5)	5(2.9)	1(3.8)
1gb1	<u>3(2.0)</u>	10(1.7)	0(4.1)
1shg	0(6.7)	9(4.1)	<u>3(5.0)</u>
1jic	0(10.0)	6(6.3)	<u>2(8.0)</u>
1r69	1(2.9)	1(2.3)	<u>2(2.8)</u>
1c9oA	5(3.2)	10(3.2)	<u>6(3.1)</u>
1mjc	<u>6(3.8)</u>	9(3.2)	3(4.6)
1fgp	0(10.7)	8(8.3)	<u>1(10.1)</u>
1ubq	<u>6(2.5)</u>	8(2.7)	4(3.4)
1oqp	<u>3(5.0)</u>	6(4.1)	2(5.3)
1btb	3(4.2)	6(3.7)	<u>6(3.4)</u>
1p9yA	0(13.9)	3(7.8)	<u>2(9.0)</u>

Table 2.5: Number of good structures $N(z_{RMSD} \leq -1)$ and minimum RMSD in Å for centres of the 10 largest clusters (second column) and 10 structures of lowest Δ_{EC} for exact EC (third column) and predicted EC (fourth column). The best method (not considering exact EC) is underlined in each row. The proteins are sorted by length.

all proteins investigated and predicted EC is more effective, too, with seven cases in which the predicted EC finds more good structures as opposed to five cases for the cluster centres.

It may be argued that the number of good structures, $N(z_{RMSD} \leq -1)$ is not a good measure here as only one cluster can be expected to be really near-native. It is, however, significant that the set selected by the predicted EC only once contains not a single good structure among its ten (and this is the case for the protein with PDB id. 1gb1, for which it has already been conceded that the correlation of exact and predicted EC is very low and which may simply be a case of failed profile prediction). The set of cluster centres on the other hand contains no good structures at all in four out of twelve cases. It thus appears that clustering could not cover the relevant parts of conformation space rather often and missed out on the good structures existing within the sampled conformations. Selection by predicted structural profiles (and also by the Rosetta score) therefore is the more reliable method.

This problem becomes even more severe for those proteins where low-resolution sampling was rather poor, namely for proteins with PDB ids. 1p9yA and 1fgp. A more technical advantage of filtering by scoring instead of by clustering is that it runs faster and is more flexible as it does not have to be rerun if the number of selected structures is to be changed.

2.2.5 Filtering of CASP8 Models

As the generation of candidate sets for longer protein sequences had turned out unsatisfactorily the method of filtering by structural profiles was additionally tested on protein models submitted to CASP8 in the server-only category. With lengths of proteins-to-be-predicted ranging from 69 to 533 amino acids not all the CASP8 proteins were longer than the ones before but considering only those 20 proteins with sequence lengths beyond 120 amino acids gave a similar picture to what will be reported here. For the 29 targets considered here (others had been rejected because of ruptured chains) between 100 and 300 models per protein were available. These

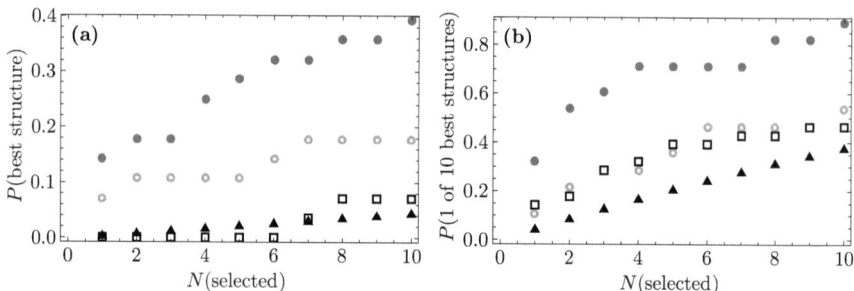

Figure 2.8: Relative frequencies of good structures for the CASP8 set, **(a)** gives the relative frequencies of choosing the best model when selecting 1 to 10 models, **(b)** gives the relative frequencies of choosing one of the ten best models when selecting 1 to 10 models. Red circles stand for selection by exact EC, orange circles for selection by predicted EC, white squares for the Rosetta score and black triangles for probabilities to find (one of) the best structure(s) in a random sample.

models were again ranked by the scoring function (2.3) based on exact EC and predicted EC or the Rosetta score (2.2). Clustering models by RMSD was also attempted but gave no significant results because of too small structure sets (only up to 300 models per protein instead of the 10,000 candidates created before). Correlation coefficients between exact and predicted EC lay between $\rho_c = 0.34$ and $\rho_c = 0.82$ and profile prediction quality was thus comparable to those from the first set.

Figure 2.8 shows the relative frequencies of choosing **(a)** the single best model when selecting between 1 and 10 models from the set and **(b)** one of the ten best models. The two parts of the figure only differ in which models are to be considered as "good" and **(a)** applies a stricter criterion. As model quality varies widely, TM-score was applied as a measure of similarity instead of RMSD values. The Rosetta energy was used as a scoring function to rank the models although the models had been created by various methods (Rosetta being only one of them). It is remarkable how well the Rosetta score performed considering models had not been optimised using this score. Nevertheless, scoring by structural profiles was even better. Again, scoring by the exact EC is clearly the best method but, as that profile would not be available for real predictions, can only be regarded as a proof of principle and of the potential of filtering by structural profiles. Filtering by the predicted EC is comparable to the Rosetta score for the laxer criterion of what is to be considered a "good" structure (Fig. 2.8 **(b)**) but substantially better for the stricter criterion (Fig. 2.8 **(a)**). This is consistent with the findings based on the candidate sets generated by Rosetta where structural profiles were better at selecting structures of minimal RMSDs.

2.3 Discussion

One of the main bottlenecks in protein structure prediction is the selection of good coarse-grained candidate structures that are worth the computational effort of refinement in detailed all-atom potentials. A promising way to select good structures is the use of so-called structural profiles that can be predicted from sequence to good accuracy.

As a proof of principle, the exact EC was used which shows the great potential of the method. Quality of predicted profiles was also sufficient to produce results comparable to or better than established methods. Notwithstanding the good results, the fact that further improvement could come from better predictions of structural profiles was shown by creating linear interpolations of exact and predicted structure profiles. Furthermore, quality of profile predictions is independent of sequence length so this part of the scheme could be applied to protein predictions of arbitrary lengths. Unfortunately, candidate sets deteriorate with sequence length quite severely and, while filtering by profiles is better than filtering by the Rosetta score for candidate sets of only moderate quality that contain at least some good structures, very poor decoy sets are beyond the reach of any of the filtering methods.

The relative performance of the predicted EC compared to the Rosetta score depends on how many structures are selected from the coarse-grained candidate sets. If fewer candidates are selected, the difference in filtering by predicted ECs or by the Rosetta score is greater and selections based on the predicted EC significantly better. This, however, diminishes with increasing selection size. It thus can be argued that filtering by predicted EC is especially useful if, due to limited computing resources, only a small number of structures is to be entered into refinement. Similarly, the strength of the method of filtering by predicted ECs is displayed for those proteins where candidate structures were only of moderate quality. While in good structure sets both methods were able to select good structures, the EC method performed better for weaker structure sets.

The general applicability to longer structures was shown using the server-only models submitted to CASP8. This test also evidences the filtering method's independence of model generation. If a stricter criterion was applied, as to which structures were to be considered "good" structures, predicted EC performs considerably better than the Rosetta score, following along the same line as the results for Rosetta-generated structure sets. Selection of conformations by exploiting information contained in structural profiles is therefore worth integrating into standard *ab inito* structure prediction methods.

3 Protein Folding Dynamics

3.1 Motivation

Even if the native structure of a protein is known it may still be unclear how this special state is attained. Most small proteins fold reversibly and it is generally accepted that the native state corresponds to the minimum in free energy [2]. Still, proteins may misfold which, if not corrected by the organism, can cause pathological aggregation of proteins [1]. Characterisation of free energy landscapes and protein dynamics may reveal intermediate states or kinetic traps that can favour such aggregation. The height of free energy barriers may explain widely varying folding rates of different proteins and conformations in states other than the native one can be investigated.

According to today's understanding of Levinthal's paradox, the potential energy governing protein folding has to be funnelled towards the native state [42, 43] to avoid sampling of an astronomically large number of conformations. A drawback of many detailed all-atom force fields is that their potentials are so complicated that the resulting energy surface is very rough. A particularly simple model, which has the advantage of a smoothly funnelled energy function, is the so-called Gō-model [12]. While early versions of Gō-models were restricted to lattices, they are now more often applied to off-lattice protein descriptions. Instead, their characteristic property is that the protein is biased toward the native structure either by making only contacts attractive that exist in the native state or by adding an energy based on deviations from native angles or distances [97, 98]. This results in a very smooth energy funnel and is based on the principle of minimum frustration [26] which sums up the notion that evolution has optimised proteins such that they may fold reliably into their respective native states.

Such simple Gō-models can be applied to relatively large proteins and may capture features related to steric effects in protein folding [99]. They are, however, incapable of reproducing phenomena such as misfolding, precisely because of the smooth energy landscape [100]. Simple models based on pairwise contacts are also insufficient at explaining two-state cooperative behaviour [101]. In practice, Gō-models are therefore often used in conjunction with additional potentials to account for e.g. hydrogen bonds or physico-chemical details of amino acids [102].

Even if the principle of minimum frustration is relaxed, native-centric models such as the Gō-model carry the assumption that the structure of the native state determines the folding behaviour. Sequence information only enters indirectly in selecting the native structure but chemical details of the sequence merely account for minor corrections to, for instance, folding rates [103–105].

In order to follow this route of dropping the principle of minimum frustration while retaining native-centricity, a protein model based on the structural profile of effective connectivity (EC) is developed. In the last chapter it was demonstrated that ECs (even inexact, i.e. predicted, ECs) were capable of selecting near-native structures from a large set of structure candidates. In this chapter the possibility of using the EC to guide protein folding to the native state is investigated. While Gō-models used in literature frequently contain terms accounting for distortion of angles or deviation from equilibrium distances in addition to the mere contact energy term, in this thesis the Gō-model is restricted to a bare minimum, i.e. contact energies. This results in a loss of resolution but is necessary to allow better comparison to the EC-model which is equally based on contact topology only. As the purpose of these investigations is not structure prediction but characterisation of folding behaviour given the native structure, the exact EC is used throughout.

This chapter is subdivided into three main parts (Sections 3.2 to 3.4) and a subsequent discussion (Section 3.5). Section 3.2 presents the model used to describe and simulate a protein and Section 3.3 gives results for successful reconstructions of a set of small proteins. Section 3.4 further analyses folding simulations, explaining the necessity of free energy landscapes (Subsection 3.4.1) and introducing relevant methods (Subsections 3.4.2 to 3.4.4). Results are presented in Subsections 3.4.5 to 3.4.7.

3.2 The Protein Model

The protein is modelled as a chain of C_α-atoms inscribed into a tube of finite thickness (with diameter 3.3 Å). Consecutive C_α-atoms are spaced 3.8 Å apart as a consequence of the rigidity of peptide bonds. The tube thickness is to account for excluded volumes of omitted side chains. Actual size of specific side chains is disregarded and uniform thickness assumed. This structure description has been adopted from Ref. [106] where it could be shown that excluded volume and hydrophobic attractive interactions alone result in typical secondary structure elements, i.e. α-helices and β-sheets [107], which are formed to achieve optimal packing of the tube and sqeeze out water [29].

The original paper [106] also includes hydrogen bonds that require specific geometric conformations but this is neither necessary for secondary structure formation nor for tertiary structure contacts and is therefore omitted here. An attractive interaction is necessary to form compact structures but instead of a hydrophobic interaction between pairs of residues I developed and use an energy based on the structural profile which also effects attraction between residues and therefore compaction of the protein. The steric energy is simplified such that overlap between different parts of the tube and too tight angles are forbidden by a very high energy penalty but apart from this no bending rigidity is involved.

In the implementation of the tube model I follow Auer et al. [108] and approximate the uniform tube by spherocylinders, cylinders capped with semispheres, whose axes coincide with the links joining consecutive C_α-atoms (see Fig. 3.1). Spherocylinders that do not share a C_α-atom are not allowed to intersect. There exists an exact method to check for tube overlaps [109] that is also more efficient than the approximation by spherocylinders used in Ref. [108] but considerably more involved. Besides, the rate-limiting step in the simulations is not the house-keeping of tube conformations but solution of the contact map's eigensystem to obtain the structural profile. Therefore the implemented version follows the approximative tube model [108].

Possible movements of the tube are the so-called pivot and crankshaft moves (see Fig. 3.2). For a pivot move a C_α-atom is picked at random, as are an axis through said atom and an angle of rotation ϕ. All C_α-atoms with indices higher than the selected one are rotated by the angle ϕ. In a crankshaft move two C_α-atoms are selected and the axis of rotation is determined by their connecting vector. The first of the two atoms is picked randomly, the second is picked randomly but at a maximum distance of five residues along the chain. Thus for C_α-atom i the partner can be $i+2$, $i+3$, $i+4$ or $i+5$ (rotation about the axis defined by the bond between i and $i+1$ would not alter the tube conformation). All C_α-atoms between the selected two are rotated by the angle ϕ about the axis of rotation.

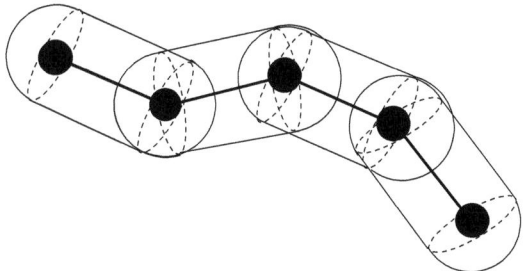

Figure 3.1: The tube model: Protein modelled as a chain of finite thickness, cylinders account for excluded volume.

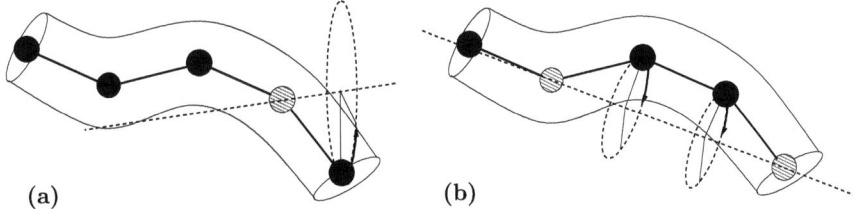

Figure 3.2: Move set of the Monte Carlo simulation, **(a)** pivot and **(b)** crankshaft move.

Values for ϕ are determined from

$$\phi = \frac{\pi}{25}\sqrt{-2\ln(r_1)}\cos(2\pi r_2), \tag{3.1}$$

where random numbers r_1 and r_2 are uniformly distributed between 0 and 1. The resulting distribution for ϕ is approximately Gaussian with mean 0 and standard deviation $\pi/25$.

These moves were used to simulate the protein in a Metropolis Monte Carlo scheme [110] where a new conformation is accepted or rejected according to the following acceptance criterion,

$$A_E(x_j|x_i) = \min\left(1, \exp\left(-\frac{E_j - E_i}{k_B T}\right)\right). \tag{3.2}$$

Here x_i denotes the old conformation and x_j the proposed new state after one of the above moves. If the new state has a lower energy E_j than the old state (energy E_i) the move is always accepted. If it means an increase in energy the new conformation is accepted with probability $\exp\left(-\frac{E_j - E_i}{k_B T}\right) < 1$. For higher temperatures T, unfavourable new conformations are thus accepted more often.

The transition probability $P(x_j|x_i)$ from state x_i to x_j consists of two factors, namely the proposition probability $g(x_j|x_i)$ determined by the move set and above acceptance probability,

$$P(x_j|x_i) = g(x_j|x_i)A_E(x_j|x_i). \tag{3.3}$$

The two possible moves were not attempted with equal probabilities but crankshaft was picked in 90 % of the cases and pivot in 10 %. This was done because the more local crankshaft move less often causes tube overlaps and the new conformation is therefore accepted more often. However, both moves are local enough to consider the Monte Carlo trajectory as pseudo-dynamic. For the same reason small values of ϕ were favoured by using above Gaussian distribution.

The Metropolis sampling scheme is a simple minimisation algorithm in that it moves towards lower energies but also overcomes barriers by occasionally accepting higher ones. But moreover, it also results in Boltzmann distributed sampling of states once the system is properly equilibrated. Equilibrium is reached if detailed balance is satisfied,

$$p(x_i)P(x_j|x_i) = p(x_j)P(x_i|x_j), \tag{3.4}$$

so there are the same number of transitions from x_i to x_j as the other way round. Then, for the Metropolis acceptance criterion (3.2) and symmetric proposition probabilities $g(x_j|x_i) = g(x_i|x_j)$ the ratio between $p(x_i)$ and $p(x_j)$ becomes

$$\frac{p(x_i)}{p(x_j)} = \frac{P(x_i|x_j)}{P(x_j|x_i)} = \frac{g(x_i|x_j)A_E(x_i|x_j)}{g(x_j|x_i)A_E(x_j|x_i)} = \frac{\min\left(1, \exp\left(-\frac{E_i - E_j}{k_B T}\right)\right)}{\min\left(1, \exp\left(-\frac{E_j - E_i}{k_B T}\right)\right)} = \exp\left(-\frac{E_i - E_j}{k_B T}\right). \tag{3.5}$$

This, together with the normalisation constraint $\sum_i p(x_i) = 1$, means Boltzmann sampling.

The symmetry of proposition probabilities and ergodicity in conformation space have to be ensured by the move set which for the above movements clearly is the case. Any state can be reached from any other state (although not necessarily in one step) and the angle of rotation is symmetrically distributed so a forward rotation is as likely as a backward rotation.

The energy of a state consists of three terms

$$E_{\text{tot}} = E_{\text{steric}} + E_{\text{EC}} + w E_{\text{SS}} \tag{3.6}$$

for the profile based EC-model and

$$E_{\text{tot}} = E_{\text{steric}} + E_{\text{Gō}} + w E_{\text{SS}} \tag{3.7}$$

for the contact map based Gō-model. The steric energy term is very simple and its only task is to prevent overlap of the tube and enforce excluded volume by a prohibitively high energy penalty. The secondary structure energy E_{SS}, with a weight w to adjust its relative strength, was introduced to prettify secondary structure elements and will be discussed below.

Both E_{EC} and $E_{\text{Gō}}$ are energy biases towards the target conformation obtained from the Protein Data Bank (PDB, [111]). The profile-based energy is defined as

$$E_{\text{EC}} = \epsilon \sum_{i=1}^{L} |t_i - c_i|, \tag{3.8}$$

where $\mathbf{t}$ is the target structure's EC-profile and $\mathbf{c}$ is that of the current conformation in the Monte Carlo (MC) simulation, L is the number of residues of the protein (or the length of the amino acid sequence). The energy unit ϵ is arbitrary and bears no obvious relation to any physical energy. The definition of Eq. (3.8) amounts to the choice of $\alpha = 1$ in Eq. (2.3) in Chapter 2. While the choice of α made no discernible difference for structure selection, this choice of $\alpha = 1$ was found to work best when trying to reconstruct a small number of protein structures. A similar cost function was also used for protein structure alignemnts for which the optimal exponent was determined as $\alpha = 1.6$ [63].

The standard contact-based Gō-energy is

$$E'_{G\bar{o}} = -\epsilon' \sum_{i,j=1}^{L} C_{ij} C_{ij}^{(t)} \qquad (3.9)$$

where C_{ij} is the current conformation's contact map and $C_{ij}^{(t)}$ the contact map of the target structure. So, whenever a native contact is formed the energy is lowered by $-\epsilon'$ and additional (undesired) non-native contacts are ignored. In the simulations for this thesis this led to too compact structures with all native but also many additional contacts. Therefore the energy was adapted and a variant Gō-model was used based on the contact overlap or ratio of native contacts q,

$$q = \frac{\sum_{i,j=1}^{L} C_{ij} C_{ij}^{(t)}}{\max\left(\sum_{i,j=1}^{L} C_{ij}, \sum_{i,j=1}^{L} C_{ij}^{(t)}\right)}. \qquad (3.10)$$

Here, the number of native contacts is divided by the maximum number of contacts in either the current or the target conformation. The contact overlap q thus does not depend on protein length which an (extensive) energy should do. This was remedied by multiplying q by residue number L to obtain the Gō-energy

$$E_{G\bar{o}} = -\tilde{\epsilon} L q \qquad (3.11)$$

which is the energy used whenever the Gō-model is discussed. Temperature T is also measured in units of ϵ or $\tilde{\epsilon}$ and k_B is set to 1 from now onwards.

Using the energy function as defined above (without a secondary structure energy E_{SS} but this makes little difference) only a few small proteins could be reconstructed [112]. At that point not only the effective connectivity was tested as a profile but also the principal eigenvector and a "revised principal eigenvector" [113], with which the problem that only information on the largest domain is contained in the PE is remedied by setting non-contacts to some small value $\delta > 0$. Irrespective of the profiles folding was very rare and inspection of protein conformations showed that they were compact but lacked typical secondary structure elements. This is because the EC-energy is not simply an attractive interaction indiscriminate of the residue but is position-specific because of the structural profile. It therefore cannot simply substitute the hydrophobic attraction of the original tube model. Introducing additional energy terms for a general hydrophobicity as in the original tube model also did not amend this as they were either too small to take sufficient effect or, if larger, interfered with the conformation of minimum energy [112].

The solution was to encourage formation of secondary structure elements in the simulations [114] by restricting the contact map, and thereby the effective connectivity, to those parts of the conformation that showed typical protein-like local structure (see Fig. 3.3). Every residue or position in the sequence was attributed a flag that indicates whether the residue

(a) 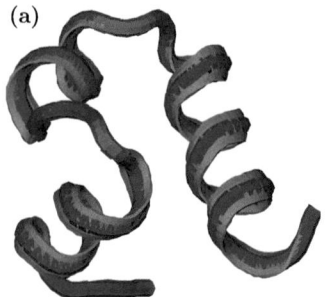 (b)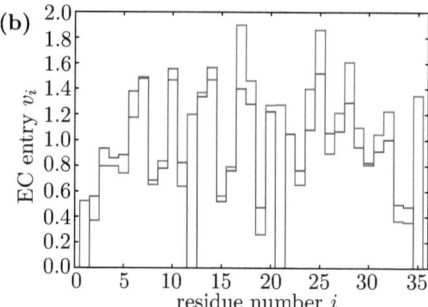

Figure 3.3: Restricted structural profiles, **(a)** backbone of the villin headpiece with protein-like residues marked blue and loops red, **(b)** full (red) and restricted (blue) EC profile of the villin headpiece.

in question is "protein-like" and all rows and columns for non-protein-like residues were deleted from the contact map before computation of the eigensystem. In the effective connectivity the corresponding entries were set to zero (see Fig. 3.3). This causes the omitted residues to be less restricted and comes at the cost of deteriorated resolution of the folded structure. This is tolerable as long as only a few positions are concerned. Usually amino acids near the ends of the sequence show less secondary structure but this is consistent with experimental data on protein structures. Single non-protein-like residues in the core are held in place by the distance constraints to their sequence neighbours as well as by steric constraints (viz. excluded volume). It is to be noted that the overall profile is not altered significantly because of this change in definition, therefore, if at some point predicted profiles are to be used for the folding, the profile's predictability should not be altered much either.

Two different criteria were explored to discern what was to be accepted as being protein-like. The idea behind both of these is that secondary structure elements are the distinct feature of proteins but whereas the first is based on local geometric properties the second relies on contact patterns. For study of the former criterion the definition of secondary structure elements given in Ref. [115] was employed, using chirality χ and distance $r_{i,i+3}$ between the end points of segments of length 4. Chirality is defined as

$$\chi = \left(\vec{e}_{i,i+1} \times \vec{e}_{i+1,i+2} \right) \cdot \vec{e}_{i+2,i+3} \tag{3.12}$$

with $\vec{e}_{i,i+1}$ being the unit vector pointing from the C_α-atom with index i to the C_α-atom with index $i+1$. For α-helices chirality is positive, $0.2 < \chi \leq 1$ and $4\,\text{Å} \leq r_{i,i+3} \leq 7.5\,\text{Å}$, for strands of β-sheets χ can be anything between -1 and 1 and $r_{i,i+3} > 7.5\,\text{Å}$. Figure 3.4 illustrates these definitions on a scatter plot of the $(\chi, r_{i,i+3})$-distribution of a representative set of proteins (PDB-select25 of May 2008 containing representative PDB structures of pairwise sequence similarity of 25% or less, [116]). If the segment matches one of the above conditions the entire segment is declared α-helical or β-sheet-like.

The contact criterion of protein likeness is based on patterns of cooperative contacts that are typical for α-helices or β-sheets (see also Fig. 1.4). For an α-helix cooperative contacts between

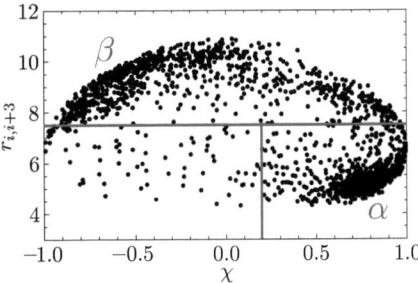

Figure 3.4: Geometric criterion of secondary structure. Segments of four C_α-atoms are considered. If chirality is larger than 0.2 and end-to-end distance is below 7.5 Å, the segment is considered helical. If end-to-end distance is above the threshold of 7.5 Å, the segment is declared part of a β-sheet.

i and $i+k$, $i+1$ and $i+1+k$ up to $i+A-1$ and $i+A-1+k$ are required. The number of residues per helix turn is 3.6 and hydrogen bonds exist between residues i and $i+4$ or, for so-called 3_{10}-helices, between residues i and $i+3$. Therefore $k=4$ or $k=3$ was allowed and the minimum size of a helix, A, was set to 4 consecutive contacts. The contact pattern of parallel β-sheets is very much the same except that $k>4$. For anti-parallel β-sheets the pattern that has to be followed is: Contact between position i and j together with contact between $i+1$ and $j-1$ etc. The minimum size of β-sheets is also set to 4 contacts. Additionally, for helices $\chi > 0.2$ was required. A third secondary structure assignment was introduced that is the "turn" which has to follow (or preceed) a helix and has the same contact pattern as a helix but not necessarily correct chirality. This was done to increase the number of residues that carry some secondary structure and thereby increase the constraints on the protein structure and ultimately increase resolution.

For target conformations, secondary structure assignments by these criteria agree fairly well with those obtained from dedicated secondary structure assignment tools such as DSSP [117] and STRIDE [118]. For only partly folded or reconstructed structures the contact criterion in particular allows considerably more distorted elements than would occur in folded structures.

Both these criteria considerably simplify finding the correct target structure, mainly by encouraging the formation of secondary structure and thus limiting the conformation space. Information about specific secondary structure elements (i.e. whether helix or sheet) from the target did not enter the simulation so neither of the criteria introduces a hidden coarse-graining to the level of secondary structure. The important difference between the two is that the contact criterion favours the formation of α-helices as their contact pattern is more local while for β-sheets distant parts of the chain have to meet, whereas the geometric criterion favours β-strands (i.e. unpaired parts of β-sheets). The latter was found to be unphysical as it resulted in strange intermediate conformations of, for example, single strands aligned to a helix although the real unit of secondary structure is the sheet and not the strand. Besides, β-sheets are indeed more difficult to fold than helices, their formation goes continously uphill in energy until the hydrogen bonds between strands are established [24]. The more realistic contact criterion was therefore used.

The $(\chi, r_{i,i+3})$-distribution found for real proteins was, however, used for the secondary structure energy E_{SS} because the secondary structure elements are usually somewhat distorted if only their contacts are prescribed and resolution of the final structure can be improved by introducing the extra energy. The histogram in χ and $r_{i,i+3}$ was determined for the PDBselect25 as of May 2008, which contains representative PDB structures of pairwise sequence similarity of 25% or less [116]. From this the free energy $F(\chi, r_{i,i+3})$ was calculated and used as potential energy $E_{SS}(\chi, r_{i,i+3})$. An entropic contribution to the abundance of helices or sheets was ignored but could be incorporated approximately as a simple offset between energies in the region associated with β-sheets and the region associated with α-helices. I verified that the total energy contribution to predominantly α- or β-proteins of comparable length was about the same and ascertained that the inclusion of E_{SS} did not change general folding behaviour but only had an effect on the resolution of the folded structure. The weight w in equations 3.6 and 3.7 was chosen as $w = 0.01$ if the energy E_{SS} was included, which meant that compact non-native conformations had comparable contributions from E_{EC} (or $E_{G\bar{o}}$) and E_{SS}. Otherwise it was simply set to $w = 0$.

I also experimented with a different secondary structure energy for simulations using the contact criterion that specifically penalised helical residues which were more distant than 6.5 Å (although still within 8.5 Å to define a contact) and rewarded those that were closer. This resulted in neater helices while only marginally changing the folding process. The energy E_{SS} described above, however, was found to be more general as it includes β-sheets and is based on distances and angles found in nature. Therefore E_{SS} was the definition of choice whenever an energy term has been used to improve the shape of secondary structure elements. Still, if not stated otherwise, neither of the terms is used in the folding simulations and $w = 0$.

There are two observations in protein folding that cannot be represented in either of these native-centric models by construction. The first phenomenon is that of cold denaturation – the native states of most known proteins are most stable around room temperature and can be denatured by either heat or cold. The unfolding transition at increasing temperature is entropy driven and can be well represented by a model in which the internal energy U takes its minimum at the target structure (by construction both $E_{EC} = 0$ and $E_{G\bar{o}} = -\tilde{\epsilon} L$ are minimal for the fully folded structure). Cold denaturation is an entirely different issue and probably a result of a change in the bulk properties of water [119] and therefore of the hydrophobic interactions. Thus, any energy terms that rely on an implicit treatment of water with interactions parametrised for conditions at which the native state is stable (of which the native-centric EC- and Gō-model are only two examples) will break down if the hydrophobic effect changes.

The other simplification of the native-centric models discussed here is that they rely entirely on the structure. Sequence information only enters very indirectly in that sequence after all does determine structure. There are, however, many sequences that fold into very similar structures and one goal of protein engineering is to design ever faster folding sequences for a given structure [120, 121]. These sequence-dependent folding rates cannot be reproduced either – only where folding rates are a consequence of the structure they might be seen in this model. Optimal folding rates are indeed strongly dependent on structure and can be predicted quite well by taking into account the number and distance of non-local contacts [103–105]. It is thus informative to test the role of the native structure in determining the folding path and rate by comparing structure-based models to experimentally observed folding behaviour.

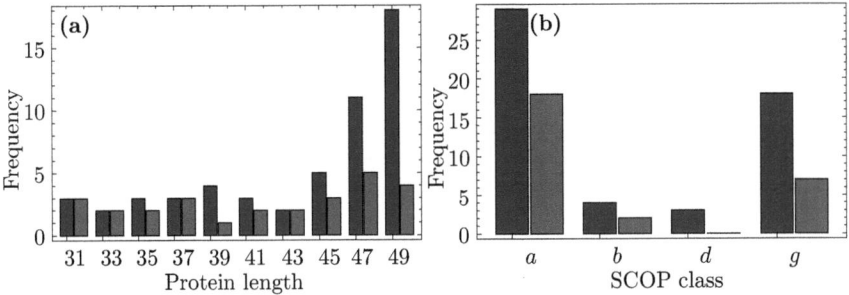

Figure 3.5: (a) Histograms of the lengths of target proteins (blue) and of reconstructed proteins (red), (b) histograms of the SCOP classes of target proteins (blue) and reconstructed proteins (red). The classes are: a (all-α), b (all-β), d ($\alpha + \beta$) and g (small).

3.3 Protein Structure Reconstruction

Before I started analysing folding trajectories or free energy landscapes I tested whether folding was successful for a representative test set of small proteins [114]. The fact that the target conformation lies in the potential energy minimum alone is not sufficient to ensure finding it in an MC simulation, additionally the energy function has to be sufficiently smooth to allow taking a pathway that will lead to the minimum. If this condition is not fullfilled and there is no funnelling towards the native state we are back at the Levinthal paradox and have to randomly try an astronomically large amount of conformations.

In order to test my model I compiled a test set from the PDB [111] of small representative proteins of length between 25 and 50 amino acids. I checked compatibility with the tube model and discarded structures with disrupted chains, atypical bond lengths and bond angles or tube overlaps. Additionally I required the structure to have at least 70% protein-like residues when applying the contact criterion described above. This left me with 1507 out of 3706 small proteins (most of the discarded protein structures did not meet the requirement on 70% amino acids with cooperative contacts) which I grouped according to their SCOP [47] (Structural Classification of Proteins) folds. For the SCOP classes a (predominantly α), b (predominantly β), d ($\alpha + \beta$) and g ("small") the longest representative of each fold was selected to serve as a target in folding simulations. SCOP class c (α/β, a special β-α-β-motif) did not occur among these small proteins and all other classes (such as membrane proteins, coiled coil or low-resolution structures) were omitted.

Figure 3.5 shows histograms of the length and SCOP class distribution of targets (blue) and successful reconstructions (red). While for most small structures the target conformation could be found in the EC-model, reconstruction of proteins longer than about 45 amino acids only succeeded in a few cases (Fig. 3.5 (a)). This is no general breakdown of the model for longer proteins but simply owed to computation time. Unsuccessful folding attempts were interrupted after about four days or $45 \cdot 10^6$ MC steps. SCOP classification also played a role in whether a structure could be reconstructed. Not only were α-proteins most abundant in the test set, they were also easiest to reconstruct with 18 out of 29 structures. Out of the four β-proteins two

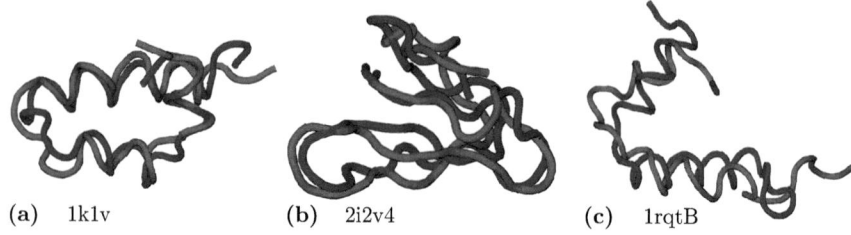

(a) 1k1v (b) 2i2v4 (c) 1rqtB

Figure 3.6: (a) Reconstruction of the protein 1k1v, RMSD to native structure is 2.4 Å, (b) reconstruction of chain 4 of the protein 2i2v with RMSD 3.0 Å and (c) reconstruction of chain B of the protein 1rqt with RMSD 7.0 Å. Target structures are blue and reconstructed structures red.

could be folded to the native state and none of the three $\alpha + \beta$-proteins (which with amino acids numbers of 48 to 50 were also among the longest structures). SCOP class g, "small proteins", contained 18 targets of which six were successfully recovered. Of these six proteins two each were all-α proteins, all-β proteins or contained both kinds of secondary structure elements. All in all, these results are not very surprising – folding gets more difficult the longer and more complex the protein structures are. The EC-model also shows realistic behaviour in kinetically favouring α-helical proteins somewhat. During folding simulations it was observed that secondary structure elements were rather stable in their assignment to one of the two kinds once they had formed and α-helices beat β-sheets in folding velocities. This observation will be discussed some more for folding of the WW domain (see Section 3.4.5).

Resolution of successful reconstructions varies over a wide range of root mean square deviation (RMSD) of C_α-atoms, the best ones having 2.2 Å and the worst 9.6 Å when measured over the entire chain. If unstructured tails are omitted RMSD improves to lie between 1.5 Å and 9.1 Å but the main issue is that there were several non-compact structures in the test set. Figure 3.6 shows a reconstructed structure for each (a) good, (b) intermediate and (c) rather poor resolution. The DNA-binding domain of protein MafG (PDB id. 1k1v) folds autonomously to a compact structure while the structure in (b) is a small subunit (chain 4) of a large complex with PDB id. 2i2v. Part (c) of Fig. 3.6 finally shows a protein domain (PDB id. 1rqt, chain B) whose structure was experimentally determined in a dimer with a copy of itself. It therefore has a rather open conformation where stabilising constraints would come from inter-chain contacts. Resolution for weakly interconnected structures is quite bad when measured in RMSD, but for comparison of the EC-model to a Gō-type model RMSD is not very informative.

For all structures observed in these reconstruction runs, identical EC-profiles mean identical contact maps corroborating the statement that the mapping from profile back to contact map is unique and no information lost in the calculation of profiles. Some resolution may be lost, however, because of the restriction to amino acids showing cooperative contacts which would not have been necessary for a Gō-model based on the contact map. In Fig. 3.7 I therefore show a histogram of contact overlap for unrestricted contact maps, q_{full}, for successful reconstructions (i.e. profile identical to target). This contact overlap q_{full} ranges from 74% to 100% which can be enough to allow reconstruction to good resolution [122, 123] under some additional assumptions on the structure's compactness.

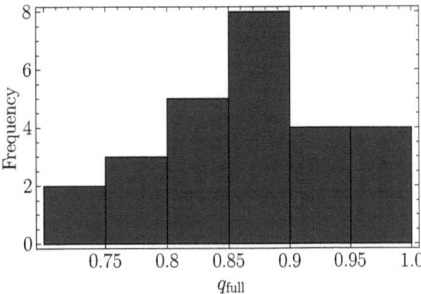

Figure 3.7: Distribution of full contact overlaps, i.e. not restricted to cooperative contacts, for all successful reconstructions.

In conclusion it is possible to reconstruct the three-dimensional structures of small proteins from their effective connectivity profiles, which therefore, to the best of our knowledge, are equivalent to the contact maps from which they had been calculated. All-α-proteins are favoured somewhat over proteins containing β-sheets and resolution is acceptable for compact protein structures. In the following I therefore concentrate on small autonomously folding proteins (or protein domains) of compact shape.

3.4 Folding Simulations and Free Energy Landscapes

This section will cover equilibrium and non-equilibrium properties of protein folding in the EC-model. Differences to the Gō-model will be worked out and, for three example proteins, compared to both experimental results and those obtained from Molecular Dynamics (MD) simulations. First, time evolution and distribution of such quantities as potential energy, RMSD and number of contacts in single simulation runs will be examined more closely. These simulations can be interpreted as specific trajectories on a free energy landscape defined by the model (either EC-model or Gō-model). A major goal therefore is to construct free energy landscapes in suitable order parameters that set the scene for folding pathways.

In principle, any property monitored during simulations can be used as order parameter to group microstates, or specific conformations, into macrostates characterised by that property. The term is borrowed from the context of phase transitions where an order parameter of 0 would describe the unordered and 1 the completely ordered phase. Here, the order parameter need not run from 0 to 1 but must be capable of distinguishing folded and unfolded state and possibly further intermediates. In using this term in this context I follow common practice as in e.g. Refs. [27, 107, 124, 125] where any observable that identifies configurations in the native and unfolded state can be used.

A reaction coordinate, on the other hand, is required to correlate with the location of the state on the folding pathway or its probability to reach the folded or unfolded state first [125]. In particular, a reaction coordinate should identify the transition state (or transition state ensemble) by grouping those configurations together that are equally likely to fold or unfold and by locating them on the free energy barrier that separates folded and unfolded state. There exist methods to identify optimal reaction coordinates [126] but these quantities are usually not easily accessible to interpretation as physical quantities. Therefore in this thesis coordinates optimally resolving, for example, the transition state are foregone and observables such as helix content used instead.

In order to construct free energy landscapes that cover a sizeable portion of phase space spanned by the order parameter(s) two methods were employed. The first method, constrained sampling, ties the simulation to a region in phase space that is to be sampled by adding a bias potential and subsequently glues different simulations together [127, 128]. This method is time-consuming because many simulations have to be run for small regions and errors in free energy differences between more distant regions accumulate. A more sophisticated method is so-called metadynamics [129] that utilises a history-dependent bias potential. Therefore, the three example proteins, namely the villin headpiece subdomain, a homologue of the peripheral subunit-binding domain (PSBD) family BBL and the WW domain, were studied in detail mainly using metadynamics to construct free energy landscapes. The results show that the EC-model agrees better with experimental evidence and observations made in MD simulations than the contact-based Gō-model does. This agrees with heat capacity curves and distributions of potential energy in the EC-model and the Gō-model in which notable differences between the Gō-model on the one and the EC-model on the other hand were confirmed.

3.4.1 Folding Simulations, Time Series and Distributions of Observables

Following the time evolution of such observables as potential energy, contact number or helix content gives a first impression about the folding mechanism and may hint at the existence of metastable intermediate states. Potential energy is a very obvious choice of order parameter as it shows how the energy minimisation proceeds, the other two quantities, contact number and helix content, are owed to the model and the protein of interest.

The villin headpiece subdomain consists of three α-helices and 27 of 35 residues are helical in the native state, so helix content monitors the amount of secondary structure formation. Nevertheless, helix content can also be an interesting observable for proteins that are predominantly built from β-sheets in the native state if there are helical (off-pathway) intermediates. Contact number is a measure of the structure's compactness and for the models investigated here, which are based on contact topology, usually more suitable than radius of gyration or end-to-end distance because of the relative freedom in inter-residue distances. The number of contacts is generally more robust and indifferent to low resolution of amino acid positions.

The villin headpiece subdomain will serve as an example to illustrate different methods and observables. A more thorough investigation and comparison to experimental or simulational results for the villin headpiece can be found in Section 3.4.5.

Figures 3.8 and 3.9 show time series for the villin headpiece in the two different models and at different temperatures. In Fig. 3.8 **(a)** the energy evolution is shown. It appears that the protein remains at an energy of about $3\,\epsilon$ before it suddenly drops to 0 and finds the native state. Energy fluctuations are obviously larger at higher temperatures ($T = 0.08\,\epsilon$ as compared to $T = 0.06\,\epsilon$) and, for low temperatures, the structure occasionally gets caught at the higher energy and does not reach the target conformation within the simulation time. Closer examination of the figure reveals a second typical non-zero energy region of about $2\,\epsilon$ that is sampled in both successful simulations. Once the native state has been reached it is very stable.

The time evolution of helix content and contact number at $T = 0.08\,\epsilon$ can be seen in Fig. 3.8 **(b)** (the red curve in **(b)** corresponds to the blue curve in **(a)**). While helix content

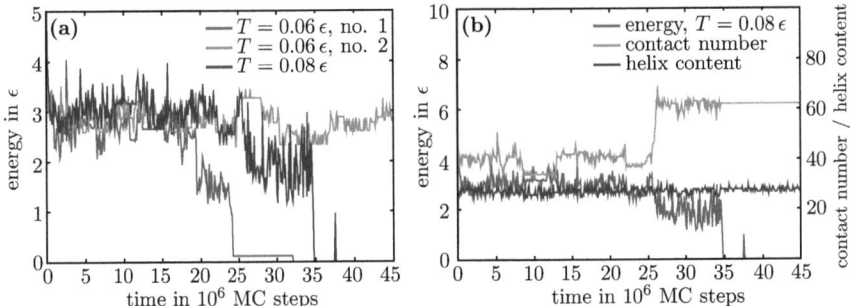

Figure 3.8: Folding time series for the villin headpiece in the EC-model. **(a)** Time evolution of energy for three different simulations at two different temperatures and **(b)** time evolution of energy, contact number and helix content for $T = 0.08\,\epsilon$ for the villin headpiece in the EC-model.

fluctuates around the target value of 27 right throughout the entire time interval, the contact number shows more interesting behaviour. It stays at a value of about 40 contacts for more than half of the simulation time, switches to the "correct" value of about 60 and is soon followed by the energy dropping to zero. Once the potential energy has reached zero, fluctuations in the contact number subside while they persist in the helix content. This is explained simply by the existence of the secondary structure assignment "turn" that allows the corresponding residues to partake in the EC profile but not be counted for helix content. It has to be noted that 40 contacts are only little more than necessary to define the helices and are mostly intra-helical contacts. Another side note is that even though assignment of "helicity" happens very early in the time series these are not fully formed and neat helices but rather deformed. Many of the contacts that also scaffold the secondary structure elements are not formed until much later.

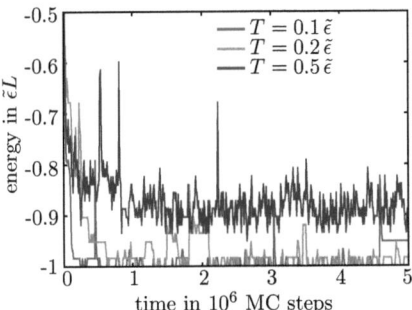

Figure 3.9: Folding time series for the villin headpiece for the villin headpiece in the Gō-model at three different temperatures.

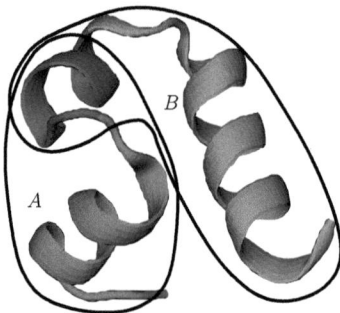

Figure 3.10: Definition of overlapping parts A (first two helices) and B (last two helices) for the villin headpiece.

In the Gō-model (Fig. 3.9) no such intermediate energies can be seen. If temperatures are low enough ($T = 0.1\tilde{\epsilon}$ and $T = 0.2\tilde{\epsilon}$) the protein folds quickly to the native state but then fluctuates around it much more than was the case in the EC-model. Size of these energy fluctuations varies and there appears to be no typical state that is visited during these deviations from the native state. If temperature is increased (e.g. $T = 0.5\tilde{\epsilon}$) the protein does not fold any more but it is important to note that no distinct transition temperature can be found.

The next step was to identify intermediate states that had been observed by other groups in the folding simulations. Lei et al. [130, 131] ran extensive MD simulations of an all-atom force field for the villin headpiece and were able to follow the entire folding process from extended conformations to the target state. They used the observables illustrated in Fig. 3.10, namely the piecewise root mean square deviations (RMSD) to the target of parts A and B, where A consists of helices 1 and 2 and B consists of helices 2 and 3. They created a free energy landscape from this and found intermediate states in it where only two of the helices had aligned correctly.

Figure 3.11 shows the joint distribution of these two observables during folding averaged over nine simulation runs each for **(a)** the EC-model and **(b)** the Gō-model. The darker the colour the more often the spot in conformation space has been visited. White regions have not been visited at all, colouring for the EC-model is held in blue and colouring for the Gō-model in red. This is not a free energy landscape as it concentrates on the transient part of the simulations and precisely not on the equilibrated part. Intermediates seen in the MD simulations could not be reproduced but another difference between EC-model and Gō-model became evident. To focus on the folding process simulations were stopped after twice the time they needed to reach the target state for the first time (or after a maximum simulation time of $45 \cdot 10^6$ MC steps). The EC-model shows two maxima in this distribution (Fig. 3.11 **(a)**), one rather sharp one at RMSDs of below $2\,\text{Å}$, which corresponds to the native state in moderate resolution, and another broad one that is centered at about $(\text{RMSD}_A, \text{RMSD}_B) \approx (3\,\text{Å}, 3.5\,\text{Å})$. In the Gō-model there is no such second maximum, instead the native state appears more blurred out which is consistent with the above observation of fluctuations. The temperature is selected such that the EC-model folds reliably, $T = 0.08\epsilon$, and the native state, defined by contact overlap $q = 1$ to the target, is still well in the centre of the free energy minimum for the Gō-model, $T = 0.1\tilde{\epsilon}$.

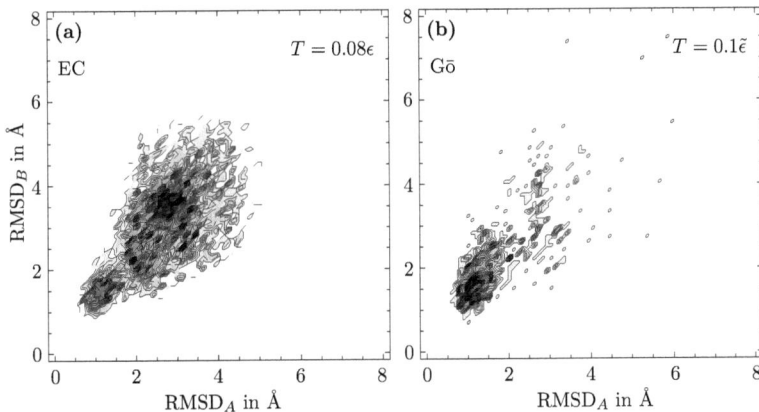

Figure 3.11: Average distribution of $RMSD_A$ and $RMSD_B$ as seen during folding of the villin headpiece in **(a)** the EC-model and **(b)** the Gō-model. The target conformation is located at (0,0) but the distribution of conformations with contact maps identical to the target one are centered at about (1.5,1.5) in these RMSD-coordinates. Darker colours (blue or red) stand for a higher number of counts. Contour lines are spaced **(a)** 2 counts or **(b)** 1 count apart.

It has to be noted that the intermediates observed in the MD simulation (but not in the EC-model) could not be experimentally confirmed. They might, therefore, be artefacts of the specific force field used. Another caveat is the low distance resolution in the model developed for this thesis which makes RMSD not a very exact observable and intermediate structures might be missed. Figure 3.12 therefore shows the distribution of the two aforementioned order parameters, helix content N_H and contact number N_C, during folding. The red rectangle indicates the location of the target structure at $(N_H, N_C) = (27, 62)$. Here, there is also a clear difference between the EC-model and the Gō-model. While the EC-model sees many structures at correct helix content but too few overall contacts this is not the case in the Gō-model which finds some structures with too little helix content but the correct number of contacts. This latter observation is probably also due to the soft criterion for secondary structure so that residues can count as β-like with only a few contacts changed compared to α-like residues. The local maxima at approximately $(N_H, N_C) = (27, 40)$ and $(26, 34)$ in the EC-model reflect the difficulties in finding the correct tertiary structure. The smaller maximum at very low helix content and high contact number $(4, 80)$ and possibly also the one at $(19, 65)$ may indicate off-pathway intermediates of alternative secondary structure.

These figures are already reminiscent of free energy landscapes and maxima in the distribution of order parameters would correspond to minima in the free energy landscape. However, to correctly assess the relative depth of different minima and the height of barriers it is necessary to run these simulations in equilibrium. For this the transient part of the simulations (i.e. the folding process) would have to be discarded. At temperatures below the folding temperature, this would mean to wait until folding is complete and the native state has been visited for the

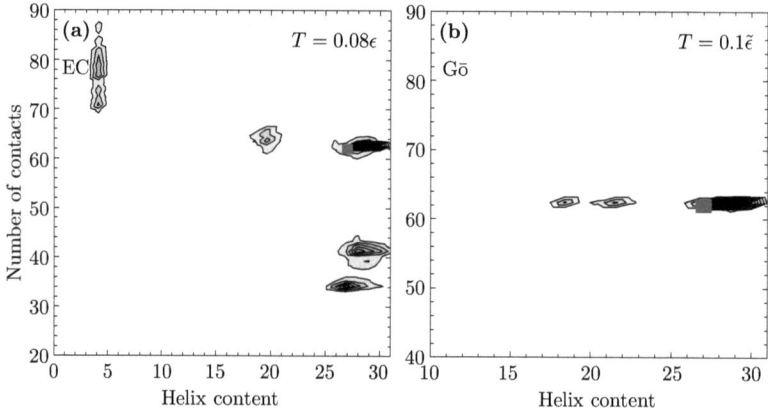

Figure 3.12: Average distribution of helix content and number of contacts as seen during folding of villin headpiece in **(a)** the EC-model and **(b)** the Gō-model. Darker colours (blue or red) stand for a higher number of counts, contour lines are spaced 10 counts apart. The red rectangle marks the location of the native state.

first time. It is, therefore, much simpler and the equilibration time much shorter when starting from the target conformation right away.

The drawback with this approach is that only conformations close to the native state will be sampled and most of the interesting regions that contain putative intermediates might not be visited. Therefore revised sampling schemes become necessary that allow the simulation to cover more of the conformation space.

3.4.2 Constrained Sampling

While the Metropolis method theoretically ensures Boltzmann weighted distributions in the long run, in practice it often gets caught in local minima and is not well-suited to compute free energy landscapes over a sizeable portion of conformation space. In biased sampling schemes the energy function of interest is replaced by a different function that is chosen such as to sample more conformations. This alters the energy dependent acceptance criterion of the standard Metropolis algorithm and allows the system to visit those regions of conformation space more frequently that would normally be ignored. Thus free energy of high-lying regions can be determined and, more importantly, the system can escape local minima to sample others which may even be of comparable depth.

The restricted free energy of an order parameter s characterising macrostates is defined as

$$F(s) = -T \ln p(s), \tag{3.13}$$

where the probability of state s, $p(s)$, is proportional to the histogram gathered during canonical (Boltzmann) sampling $N(s)$. Thus, up to a constant, the free energy equals

$$F(s) = -T \ln N(s). \tag{3.14}$$

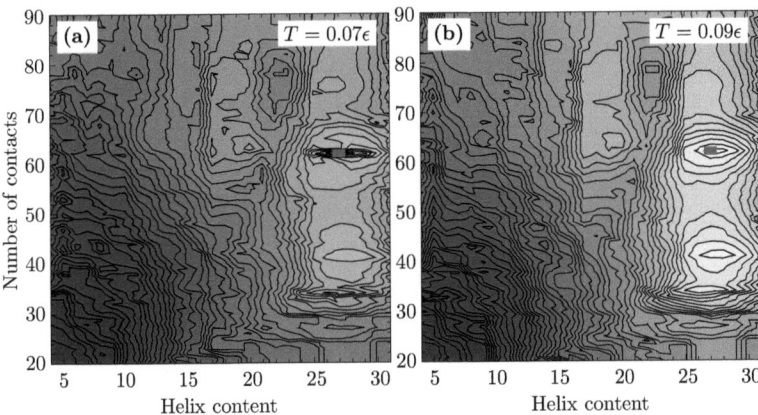

Figure 3.13: Constrained sampling for villin headpiece for two different temperatures in the EC-model with order parameters helix content and number of contacts. Contour lines are spaced at ΔF-intervals of $0.1\,\epsilon$, large values of F are blue and low values white. The red rectangle marks the location of the native state.

If the state s, however, is never sampled no statement can be made about the free energy at that position.

A simple method to enhance sampling of those regions is to divide the conformation space into smaller overlapping regions and restrict the simulation to those parts by adding a prohibitively high energy bias to all conformations outside [128]. Thus, free energy differences can be determined within all the small regions and, thanks to the overlap, added up to create a free energy landscape for the entire conformation space. This method is a special case of umbrella sampling [27, 127, 132] where more general biasing potentials are used to bridge conformation space between two regions of interest.

Figure 3.13 shows two examples of such free energy landscapes at different temperatures. Here s is two-dimensional and consists of helix content and number of contacts, the order parameters used before. To obtain these landscapes the space spanned by N_H and N_C has been partitioned into regular tiles of size 3×7 with an overlap of 1 (N_H) and 2 (N_C) to either side. Simulations were started from the native state and counting for $N(s)$ started once the system entered the allowed tile. Free energy differences were then calculated within the individual tiles and offsets determined by minimising the overall error (squared distance) between free energies from neighbouring tiles. Each (inner) tile has 8 neighbours (direct neighbours and diagonal neighbours), edge or corner tiles have correspondingly fewer neighbours. For M tiles $M-1$ offsets are determined. Again, a constant might be added to the free energy and the choice of which tile is left unshifted is arbitrary.

The colour coding in Fig. 3.13 is such that regions of high free energy are depicted in blue and free energy minima in white. The free energy difference between successive contour lines is $\Delta F = 0.1\,\epsilon$. The minimum free energy is adopted in Fig. 3.13 **(b)** at $(N_H, N_C) = (27, 62)$. The absolute value of the free energy at the minimum is not calibrated to any reference energy and carries no information (as reflected in the arbitrary choice of which tile is unshifted) and only free energy differences are of interest.

The colour scheme is inverted when compared to that of Fig. 3.12 **(a)** where regions of higher sampling probabilities are darker. Again, a red mark identifies the position of the native state and is situated in the free energy minimum for both temperatures. The maxima in the joint distribution from folding simulations centered at approximately $(N_H, N_C) = (26, 34)$ and $(N_H, N_C) = (27, 40)$ can be found as local minima in free energy as well. The position of the free energy minimum at lower helix content, $(19, 65)$, exists, too, but is not so easily discernible. For higher temperatures the free energy minimum containing the native state and those containing conformations of less helix content broaden and the local minima grow deeper and compete with the native state minimum ($T = 0.09\epsilon$, Fig. 3.13 **(b)**). The region around $(19, 65)$ becomes a more apparent local minimum and is separated from the global minimum by a small barrier accounting for the maximum at $(19, 65)$ seen during folding (Fig. 3.12 **(a)**). Other small minima are probably spurious and the free energy landscape far from the native state, i.e the region of low helix content and low contact number, is not very reliable.

There are two disadvantages connected to this type of sampling. The first one is that errors in the free energy offsets add up and free energy differences between distant areas in the landscape are therefore not very precise. Another, actually more severe, problem is the question of how to choose the size of tiles. If tiles are too large, the result is the same as that for the entire conformation space and some regions will not be sampled. This also means that shifts between neighbouring tiles cannot be determined and the landscape becomes disconnected. If, on the other hand, tiles are too small, movement of the chain is severely restricted and possibly not ergodic (within the tiles) anymore as some conformations that are mapped onto the same order parameter cannot be reached from other conformations within the same tile. This problem may only be remedied by running a large number of simulations that sample different conformations of the same order parameter and averaging over them.

Another possible issue, viz. depletion near the edges of a tile, plays no role in this context. All microstates still connect to the same number of conformations even if these are very high in energy and never visited. What is lost in influx from other conformations is gained by an increased probability of staying in place when unfavourable neighbouring conformations are attempted.

All in all, this method is a very time-consuming one as many simulations have to be run for individual tiles but it is suitable to ensure sampling of the entire conformation space and thus map out large free energy landscapes.

3.4.3 Metadynamics

An algorithm useful for global optimisation is the so-called "Energy Landscape Paving" [133] where conformations x are grouped into macrostates by some suitable order parameter s and macrostates that have been explored previously are discouraged in further steps by adding a time-dependent bias potential to the original energy function $E(x)$,

$$E'(x,t) = E(x) + V(s(x), t). \tag{3.15}$$

The bias potential $V(s, t)$ depends on the number of times the order parameter s has been visited before,

$$V(s,t) = \alpha \ln(1 + N(s,t)), \tag{3.16}$$

where the histogram $N(s,t)$ is updated after every MC step. Here it is assumed that orthogonal degrees of freedom equilibrate faster and conformations of identical s are properly sampled before moving on. Local minima in the free energy landscape are thus filled up eventually and the system escapes the minimum, then reaches the next one which it will also escape after some time and so forth. The weight α determines the filling rate of minima and, by changing it, the time the system spends in local minima can be tuned. Small values for α give the system enough time to equilibrate conformations of equal s before moving on but also cause an increase in simulation time.

This method can not only be used for minimisation but histograms collected from this sampling (in spite of the time-dependent biasing potential) can also be used to calculate free energies [129], a method dubbed (well-tempered) metadynamics by the authors of Ref. [129]. There, the bias potential $V(s,t)$ is defined as

$$V(s,t) = \Delta T \ln\left(1 + \frac{\omega N(s,t)}{\Delta T}\right) \tag{3.17}$$

with two tunable parameters, ΔT and ω. For this thesis the method was adjusted by setting $\alpha = \Delta T = \omega$ such that Eq. (3.17) simplifies to Eq. (3.16).

Thus, the time-dependent energy $E'(x,t)$ is used instead of the energy term $E(x)$ in Metropolis sampling. As this energy, over $N(s,t)$, depends on simulation history the sampling trajectory no longer constitutes a Markov chain. Still, it is possible to extract (equilibrium) free energy information from such a simulation. Once the histogram $N(s,t)$ does not change much in relative terms, $N(s,t+1)/N(s,t) \approx 1$, detailed balance, Eq. (3.4), is approximately satisfied

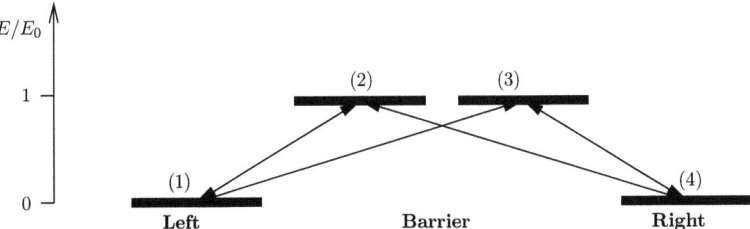

Figure 3.14: Model system consisting of four microstates that are grouped into three macrostates **Left**, **Barrier** and **Right**.

	Microstates x					Macrostates s				
T/E_0	0.01	0.1	0.3	0.5	1	0.01	0.1	0.3	0.5	1
$\Delta F/E_0$ exact	1	1	1	1	1	0.993	0.931	0.792	0.653	0.307
$\Delta F/E_0$ no bias	-	1.026	1.001	1.002	1.009	-	0.958	0.804	0.662	0.323
$\Delta F/E_0$ bias	1.000	1.000	1.005	1.005	1.005	0.993	0.932	0.799	0.663	0.320

Table 3.1: Free energy difference between barrier states and minima. For microstates x the free energy F is averaged over states (1) and (4) and states (2) and (3), whereas for macrostates s the free energy F is averaged over **Left** and **Right**.

locally and at any moment in time. We can therefore write the instantaneous probability density, with E' as sampling energy, as

$$p(s,t)\big|_{E'} = \exp\left(-\frac{F(s)+V(s,t)}{T}\right) = \exp\left(-\frac{U(s)+V(s,t)-TS(s)}{T}\right), \quad (3.18)$$

where $F(s)$ is the free energy of interest. This expression is analogous in form to

$$p(s)\big|_E = \exp\left(-\frac{F(s)}{T}\right) = \exp\left(-\frac{U(s)-TS(s)}{T}\right), \quad (3.19)$$

with the bias potential $V(s,t)$ augmenting the internal energy $U(s)$. The entropy term $S(s)$ is of course left unchanged by the change in sampling.

For large times, $V(s,t)$ varies only slowly and $p(s,t)$ will do likewise. This in turn means that $N(s,t)$, the histogram gathered up until t, will become proportional to $p(s,t)$ for long enough times. For $N(s,t)$ growing proportionally, $V(s,t) = \alpha \ln(1 + N(s,t))$ will converge modulo a (time-dependent but s-independent) offset which yields

$$\exp\left(-\frac{F(s_i)+V(s_i,t)-F(s_j)-V(s_j,t)}{T}\right) = \frac{N(s_i,t)}{N(s_j,t)}, \quad (3.20)$$

in analogy to the unbiased sampling of Eq. (3.5). On the other hand, for large N

$$V(s,t) = \alpha \ln(1+N(s,t)) \approx \alpha \ln N(s,t). \quad (3.21)$$

Substituting this into the above Eq. (3.20) and solving for $F(s_i) - F(s_j)$ results in

$$F(s_i) - F(s_j) = -(T+\alpha) \ln \frac{N(s_i,t)}{N(s_j,t)}\bigg|_{t\to\infty} \quad (3.22)$$

or by setting an arbitrary reference energy

$$F(s_i) = -(T+\alpha) \ln N(s_i,t)\big|_{t\to\infty}. \quad (3.23)$$

I illustrate this method on a minimal system consisting of four microstates $x \in \{1,2,3,4\}$ where the free energy can be calculated analytically. Figure 3.14 shows those four states and the possible transitions between them. States (2) and (3) have energy $E(2) = E(3) = E_0$ and thus constitute a barrier between states (1) and (4) which have energy $E(1) = E(4) = 0$. Allowed transitions are from state 1 or 4 to state 2 or 3 and back. All proposition probabilities are $1/2$ and the condition of symmetric $g(x_i|x_j)$ is obviously fullfilled.

There are two possibilities to analyse this system, either distinguishing all the microstates x or grouping state (2) and (3) together and using the order parameter $s \in \{\text{Left}, \text{Barrier}, \text{Right}\}$. In the first case, entropy disappears and $F(x) = U(x) = E(x)$ for all temperatures. In the second case, $s = \text{Barrier}$ has a higher entropy, $S(\text{Barrier}) = \ln 2$, while $S(\text{Left}) = S(\text{Right}) = 0$ and $F(s) = U(s) - TS(s) = \langle E(x) \rangle_s - TS(s)$ so the free energy difference ΔF between barrier states and minima is $\Delta F = E_0 - T \ln 2$.

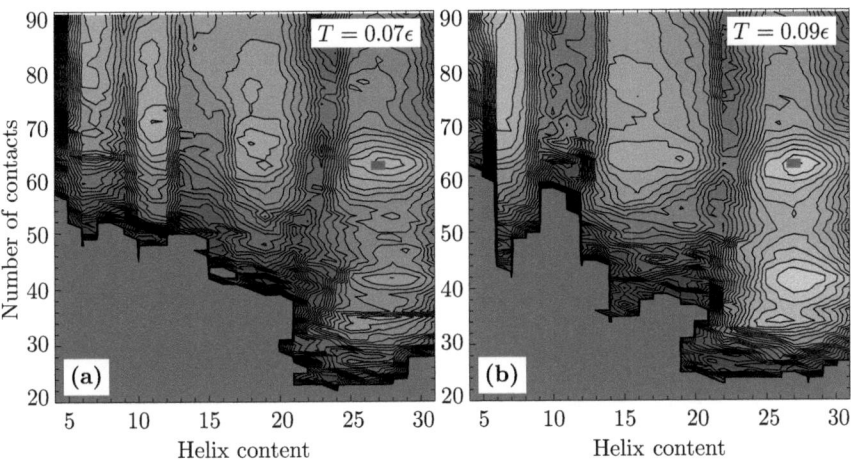

Figure 3.15: Metadynamics sampling for villin headpiece for two different temperatures in the EC-model. Temperatures are the same as in Fig. 3.13. Darker blue corresponds to higher free energy, paler colour to low free energy and white would correspond to the free energy minimum at $T = 0.1\,\epsilon$. Grey regions have not been sampled at all. Contour lines are spaced at ΔF-intervals of $0.1\,\epsilon$ and the red rectangle marks the location of the native state.

I ran simulations for both cases using 10^5 MC steps (simulations have to be run anew because the choice of order parameter determines the histogram and therefore the bias potential) and for both the agreement to the analytical result was satisfactory. Results for the free energy difference are given in Table 3.1. Even for this simple system the advantage of the biased method over unbiased sampling becomes evident for low temperatures. At temperature $T = 0.01\,E_0$ (with an initial microstate of (1)) the barrier was never climbed during the simulation time without the aid of the biasing potential and no free energy difference could be determined.

In the next step the method was applied to the protein EC-model which is in good agreement with constrained sampling, while being far less costly in terms of computation time (see Fig. 3.15, compare to Fig. 3.13). Colour coding again is such that white stands for minimal free energy (at a reference temperature $T = 1\,\epsilon$, $F = -3.9\,\epsilon$), dark blue for $F = 0$ (sampled exactly once) while grey has never been sampled and thus $F > 0$. Contour lines are spaced at $\Delta F = 0.1\,\epsilon$. The offset (for the region that has been sampled at all) is again arbitrary, for instance doubling the simulation time, and thus the number of counts, would shift F by a constant of $-T\ln(2)$. There is also moderate quantitative agreement for the free energy landscapes obtained by either constrained sampling or metadynamics, e.g. the relative depths of the local minima at lower contact number and (near-)native helix content compared to the native state minimum is $\Delta F \approx 0.3\,\epsilon$ at $T = 0.07\,\epsilon$ and $\Delta F \approx 0.1\,\epsilon$ at $T = 0.09\,\epsilon$ for both methods. The shape of the minimum centered at $(N_H, N_C) \approx (19, 65)$ agrees only qualitatively for the two methods and in metadynamics there emerges a local minimum at very low helix content and

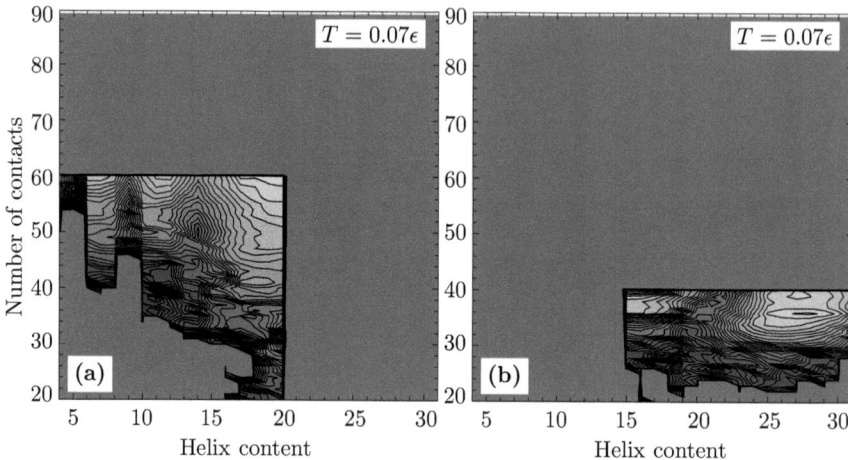

Figure 3.16: Constrained metadynamics for villin headpiece in the EC-model at $T = 0.07\,\epsilon$. In part **(a)** the sampling is restricted to the region with helix content between 0 and 20 and contact number between 0 and 60, in part **(b)** to the region with helix content between 15 and 40 and contact number between 0 and 40. Contour lines are spaced $0.1\,\epsilon$ apart but absolute values of F are un-calibrated and need to be shifted to match free energies as in Fig. 3.15.

high contact numbers nicely explaining the corresponding maximum in Fig. 3.12 **(a)** which had not been observed in constrained sampling.

3.4.4 Constrained Sampling and Metadynamics Combined

An obvious idea for improvement of the methods of constrained sampling and metadynamics is to combine the two of them. Thanks to the metadynamics method the regions of constrained sampling can be made bigger thus mitigating the problem of broken ergodicity. Metadynamics on its own already covers a sizeable portion of conformation space, however, the region of both low helix content and low contact number that has to be passed in folding simulations could be improved by some additional sampling. Therefore, sampling was restricted to one region with helix content between 0 and 20 and contact number between 0 and 60 and another with helix content between 15 and 40 and contact number between 0 and 40. Metadynamics simulations were run for each of the two regions (see Fig. 3.16). As can be gathered from the figure the system still clung close to the edges of the region that had already been sampled before. The free energy from the two new regions was subsequently shifted to minimise the error (squared distance) for overlapping parts (see Fig. 3.17).

Free energies for overlapping regions agree well (up to a constant) so the different samplings can be overlaid nicely. One part where the two constrained samplings Fig. 3.16 **(a)** and **(b)** disagree is the region with helix content between 15 and 20 and contact number just below 40

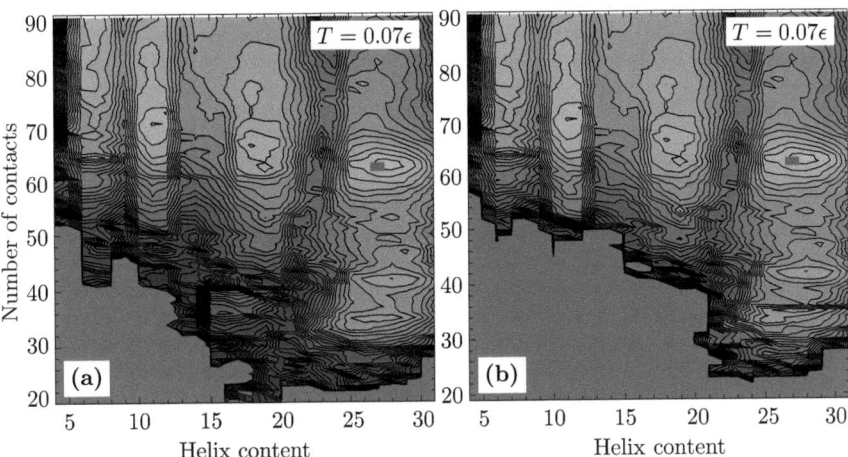

Figure 3.17: Combination of the different regions of constrained metadynamics for the villin headpiece in the EC-model at $T = 0.07\,\epsilon$. Part **(a)** shows the inclusion of the two constrained sampling simulations into the unconstrained simulation, part **(b)** repeats the unconstrained sampling from Fig. 3.15 (a) for comparison. Colour coding and contour line spacing are the same as in Fig. 3.15.

where one plot shows a rising and the other a falling flank. This disagreement is also reflected in Fig. 3.17 (a) in the form of a sharp edge where different samplings have been glued together. All in all, larger regions of conformation space may be sampled by this combined method but in order to bridge the gap to the completely unfolded state many more constrained samplings would be necessary. However, in the following examples sampling by normal metadynamics was sufficient to recover distinctive observations from experiment or MD simulations. Thus, the simpler version and not the combination of metadynamics with constrained sampling was used throughout.

3.4.5 Example Proteins and Comparison of Models

In the following subsections the method of metadynamics will be applied to three small proteins (or protein domains) that have been in the focus of experimental and computational interest for a while, the villin headpiece subdomain (which has already been used to illustrate the sampling methods in the previous sections), a homologue of the peripheral subunit-binding domain (PSBD) family which goes by the name of BBL and the so-called WW domain. Free energy landscapes will be determined for both the EC- and the Gō-model and compared to results from experiments or MD simulations with respect to the implications on folding mechanisms. Thereby it will be shown that the slightly more complicated EC-model can account for details of protein folding that the Gō-model is unable to explain.

The energy term E_{SS} was tested for these examples with its weight w from Eq. (3.6) set to 0.01. However, as this only resulted in smoother secondary structure elements and had no notable effect on the free energy landscapes or folding behaviour the energy term E_{SS} was dismissed again in order to keep the model as simple as possible. For all results shown in the following sections the weight had been set to $w = 0$.

Folding of the Villin Headpiece Subdomain

As a first example protein the villin headpiece subdomain was chosen, a structure consisting of about 35 amino acids (depending on the species and on the particular study) that attracted a lot of interest as the smallest naturally occurring polypeptide showing autonomous folding without any disulfide bonds [134]. It is involved in actin-binding [134] and consists of three fast-folding helices (see Fig. 3.18). Because of its fast folding and simple topology it is readily accessible not only to the EC-model that favours α-helices but also to MD force fields that have difficulties with β-sheets as well. It is therefore often used as a testing ground for theories and simulation methods [135–137]. The specific PDB structure that was used in the simulations for this thesis has PDB id. 1und, consists of 36 amino acids and is a subdomain of human advillin [138]. Significant secondary structure content in the denatured state of the villin headpiece subdomain was observed experimentally [139, 140] as well as in MD simulations with explicit solvent [141]. While Ref. [139] sees folding behaviour in accordance with the diffusion-collision model [142] where secondary structure elements form first and then try to assemble into tertiary structure, others stress the importance of long-range contacts and three-body correlations in stabilising secondary structure [140, 143]. Biphasic kinetics with folding happening in two stages were found experimentally [144] and in long (200 ns replica exchange and 1.0 μs conventional) MD simulations with implicit solvent [130, 131]. Those simulations showed distinct intermediates depending on which of the two helix bundles formed first.

The villin headpiece was used to test and illustrate the sampling methods in this thesis, which is why some of the results on this structure have been mentioned in passing in former sections. In this section, those results will be elaborated in more detail.

Figure 3.19 shows the free energy landscape of the villin headpiece in **(a)** the EC- and **(b)** the Gō-model. Order parameters are, again, helix content and the number of contacts which

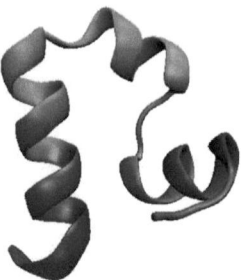

Figure 3.18: Target structure of the villin headpiece consisting of three helices. The colour scheme runs from red at the N-terminus to blue at the C-terminus.

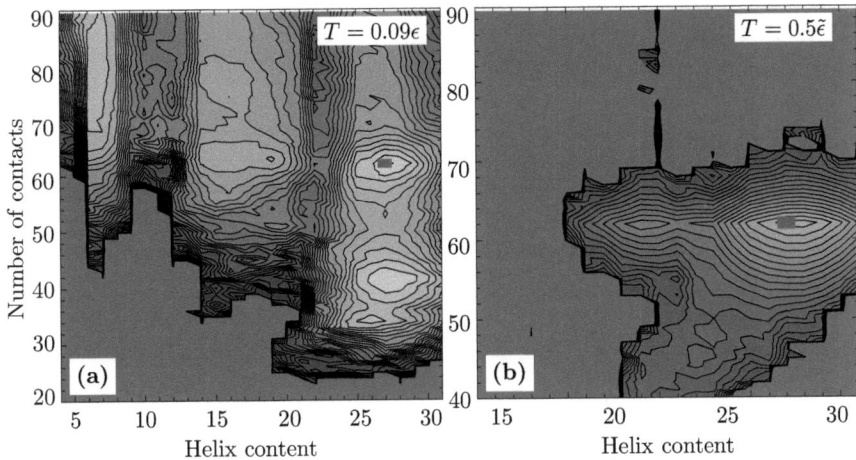

Figure 3.19: Comparison of the free energy landscape for villin headpiece in **(a)** the EC- and **(b)** the Gō-model. Order parameters are helix content and number of contacts. Note that the plot region for the Gō-model is smaller. Colour coding and contour line spacing in **(a)** are the same as in Fig. 3.15. Colour coding in **(b)** is such that white colour would be attained at the free energy minimum at $T = 0.8\,\tilde{\epsilon}$ and contour line spacing is $\Delta F = 0.5\,\tilde{\epsilon}$.

reveal several local minima in the EC-model (Fig. 3.19 **(a)**). There are two local minima close to each other at correct helix content of about 26 or 27, but with fewer contacts than the target conformation, namely 34 and 40 instead of 62. These minima correspond to structures where secondary structure has already rudimentarily formed while contacts defining the tertiary structure cannot be found so the protein remains in rather extended conformations. Another large but shallow minimum exists for too low helix content (approximately 15 to 20) and contact numbers ranging from 60 to 65. These conformations lack some of the contacts necessary to define helices but succeeded in forming some of the tertiary contacts. As it is very unfavourable for the protein in the restricted EC-model to have no secondary structure assignment at all, this will be avoided by some residues carrying the contact pattern of β-sheets instead. Another local minimum that is separated by a broad free energy barrier and therefore probably not accessible from the other minima is located at very low helix content and many more contacts (70 up to 90).

The Gō-model's free energy landscape (Fig. 3.19 **(b)**) has fewer features. The free energy landscape is largely funnelled towards the native state at $(N_H, N_C) = (27, 62)$ with only a small local minimum at $(21, 62)$. Villin headpiece has been experimentally observed to show two-state folding between the native and the denatured state so one (or possibly more) non-native states should be visible in the free energy plot. The free energy landscape in the Gō-model would instead cause downhill folding that is not limited by any free energy barrier. The colour scheme for the Gō-model has been defined such that red corresponds to $F = 0$, regions that have not

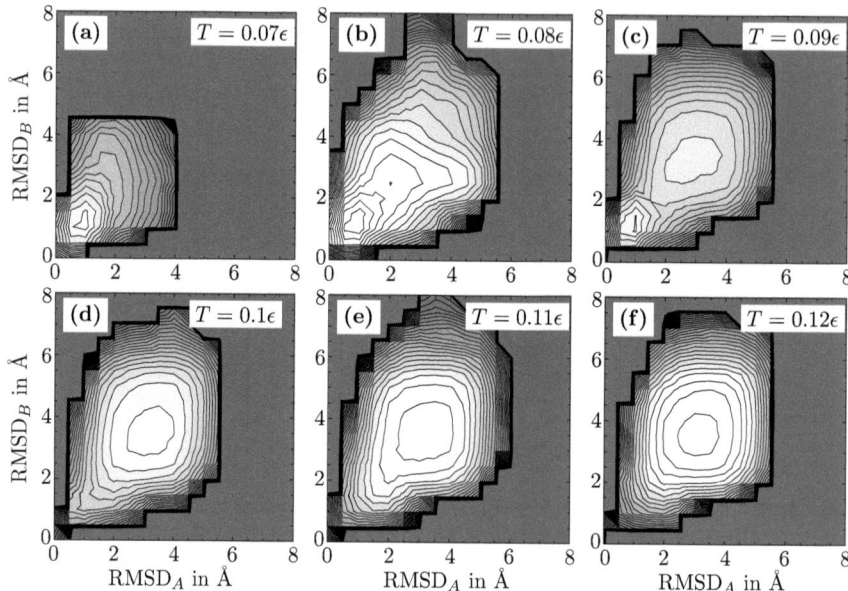

Figure 3.20: Free energy landscape for villin headpiece in the EC-model with order parameters RMSD_A and RMSD_B. For increasing temperature a new free energy minimum appears at larger RMSDs and drains the old native state minimum. Colour coding and contour line spacing are the same as in Fig. 3.15.

been sampled ($F > 0$) are rendered in grey and minimum free energy (at $T = 0.8\,\tilde{\epsilon}$, $F = -14\,\tilde{\epsilon}$) in white. Contour spacing is $\Delta F = 0.5\,\tilde{\epsilon}$.

If the free energy landscape is projected on a few order parameters only, there is a possibility that interesting details are hidden by an unfortunate choice of order parameters [145]. The denatured state of two-state folding, for example, could be inhomogeneous in helix content and thus not resolved in Fig. 3.19 **(b)**. Therefore different coordinates were tested and, in Fig. 3.20, the resulting free energy landscape is shown, spanned by order parameters RMSD_A and RMSD_B as defined in Section 3.4.1 and Fig. 3.10. The plots follow the free energy landscape for increasing temperatures ($T = 0.07\,\epsilon$ to $T = 0.12\,\epsilon$) and clearly show a transition with the maximum population shifting from the native state at $(\mathrm{RMSD}_A, \mathrm{RMSD}_B) = (1.5\,\text{Å}, 1.5\,\text{Å})$ to the denatured state at about $(3\,\text{Å}, 3.5\,\text{Å})$. It is important to note that these minima coexist for a small range in temperature ($T = 0.08\,\epsilon$ and $T = 0.09\,\epsilon$) before one or the other becomes dominant.

In the Gō-model the behaviour is fundamentally different. Instead of a new minimum growing and overtaking the old one at some transition temperature, the free energy minimum moves continuously towards higher RMSD-values. This means that there is no cooperative transition from folded to denatured state but rather a slow shift in the favoured structural ensemble. For higher temperatures the minimum of the free energy $F = U - TS$ also grows deeper. Thus, in order to have good contrast in free energy landscapes at both low and high temperatures, I

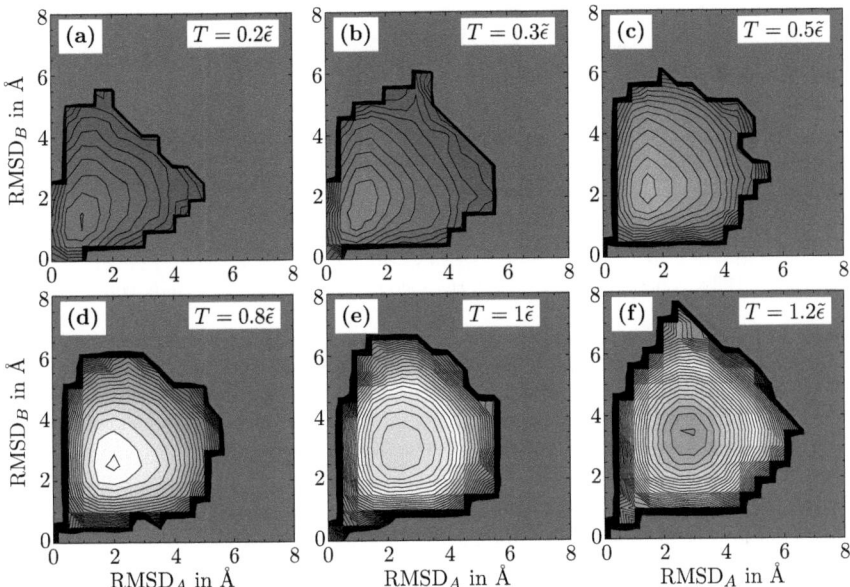

Figure 3.21: Free energy landscape for the villin headpiece in the Gō-model with order parameters $RMSD_A$ and $RMSD_B$. For increasing temperature the minimum wanders to larger RMSDs and grows ever deeper. For better contrast, the colour code (and contour line spacing) are as in Fig. 3.19 **(b)** but extended such that white still indicates the free energy value at $T = 0.8\tilde{\epsilon}$ but for higher temperatures additional colours (yellow, then green) are included. Contour line spacing is $\Delta F = 0.5\,\tilde{\epsilon}$, as in Fig. 3.19 **(b)**.

adapted the colour scheme. White colour still represents the free energy minimum at $T = 0.8\,\tilde{\epsilon}$ ($F = -14\,\tilde{\epsilon}$), red corresponds to $F = 0$ and grey to $F > 0$, but now yellow and green have been introduced for even lower free energies. Contour lines are still spaced at intervals of $\Delta F = 0.5\,\tilde{\epsilon}$.

There still is the possibility for RMSD coordinates to hide some intermediate states in the free energy landscape. In particular, these order parameters make no statement about the structure of denatured states, which was better resolved in helix content and contact number coordinates. Everything non-native will be grouped together in the region of higher RMSDs. But this is also the strength of these order parameters as they differentiate between native (low RMSDs) and non-native (high RMSDs) conformations. As there never appear two coexisting minima in the Gō-model, folding can be inferred to occur on a one-state free energy landscape without significant barriers. A similar observation was made by Piana et al. [146] who found that folding of the villin headpiece subdomain took place on a landscape of small dimensionality, which could be described by only a few order parameters, when using all-atom explicit solvent MD simulations, but not when using a coarse-grained Gō-model. My findings agree with their

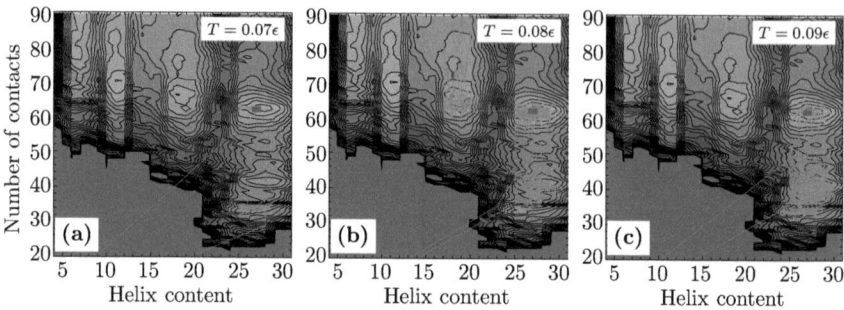

Figure 3.22: Free energy landscape for villin headpiece in the EC-model with helix content and number of contacts as order parameters. Free energy landscapes **(a)** and **(c)** are the same as in Fig. 3.15 ($T = 0.07\,\epsilon$, $T = 0.09\,\epsilon$), part **(b)** shows an intermediate temperature of $T = 0.08\epsilon$. Additionally, some folding trajectories (green and orange) have been projected onto the free energy landscapes.

results on the Gō-model but moreover show that the similarly coarse-grained EC-model succeeds in discerning distinctive states on a low-dimensional free energy landscape.

These free energy landscapes set the scene for folding transitions, and the landscape in the EC-model with helix content and contact number as order parameters displays many local free energy minima. In order to illustrate the folding process and link the equilibrium property of free energy to the non-equilibrium folding process, several typical folding trajectories were projected onto the free energy landscapes (see Fig. 3.22). Free energy landscapes and trajectories at different temperatures are shown. At low temperature, $T = 0.07\,\epsilon$ (Fig. 3.22 **(a)**), the local minima at $(26, 34)$ and $(27, 40)$ act as kinetic traps and trajectories visiting these minima do not escape to reach the global minimum within the simulation time. The free energy minimum containing the native state is either found quickly on a direct route or not at all. At $T = 0.08\,\epsilon$ (Fig. 3.22 **(b)**) the system is able to travel between the minima and also visits the minimum at lower helix content at approximately $(19, 65)$ (yellow trajectory). For one simulation (data not shown) the system directly ran into the local minimum at very low helix content and did not escape again. For $T = 0.09\,\epsilon$ (Fig. 3.22 **(c)**), the minimum containing the native state and the other two at lower contact number were still visited but population probability shifted toward the former local minima that are now of comparable depth.

Free energy landscapes and folding trajectories thus nicely agree and complement each other in that the trajectories contain temporal information and the free energy landscapes give averaged information for a larger region of conformation space. It is also reassuring that the trajectories make relatively small steps on the free energy landscapes and do not jump into distant regions (except for the very first step coming from $(N_H, N_C) = (0, 0)$). Otherwise, either the choice of order parameters or the moveset would have been dubious and with that the assumption that orthogonal degrees of freedom equilibrated quickly compared to the dynamics of order parameters, which had been necessary to justify the method of metadynamics.

The first jump occurring from the extended chain can be explained by the quick collapse to more compact (though assumably unfolded or incorrectly folded) structures. The completely unfolded, extended chain itself therefore is not a typical structure of the denatured state which is probably better represented by the local minima around $(N_H, N_C) = (27, 40)$. This hypothesis is confirmed by the observation that population of this macrostate of correct helix content but too few contacts goes up for increasing temperature (Fig. 3.22 (c)).

If the time resolution is improved and every MC step plotted for the first 50,000 steps this jump also decomposes into many small moves. In this context it is also interesting to investigate at which point the two low-temperature trajectories, started with different seeds and reaching different minima (Fig. 3.22 (a)), separate. According to the simulations in higher time resolution this happens quite early on, in the region of both low helix content and low contact number that, unfortunately, has not been sampled in the metadynamics simulations producing the free energy landscape. It is therefore unclear whether there exists a saddle point or whether trajectories just separate on an overall descending flank, simply because different conformations are proposed in the simulation.

The intermediate states observed in Ref. [130] could not be found in the simulations of the EC-model and the presentation of Fig. 3.22 does not show biphasic kinetics. Particularly the latter difference could quite simply be a result of the order parameters chosen. Going back to the time evolution in Fig. 3.8 (b), after the rise in contact number at a "time" of $25 \cdot 10^6$ MC steps fluctuations cease at a later point, that is, after about $35 \cdot 10^6$ MC steps. This is accompanied by a drop in potential energy revealing a second stage of folding. A more thorough investigation of folding would have to go beyond free energy landscapes in predefined order parameters and analyse kinetic clusters [147, 148], in my case probably best based on contact maps as microstates.

For the villin headpiece, in all order parameters investigated, the EC-model can account for experimentally observed two-state folding while the Gō-model cannot. The residual helix content in the unfolded state observed in experiments can also only be explained in the EC-model (but not in the Gō-model) where the denatured state appears to be already largely helical.

Folding of the Protein BBL

BBL is an independently folding helical domain consisting of around 45 residues (depending on the species and where the domain's limits are set). It is involved in binding to a specific enzyme and is part of the core of a large multienzyme complex responsible for glycolysis [149]. Its folding behaviour has been discussed controversially in the literature, in particular as a possible example of barrierless downhill folding in nature. Downhill folding without a rate-limiting barrier and only a single free energy minimum is not unusual in model systems (in fact, purely additive models such as the Gō-model frequently show it) but it is an open question whether this mechanism is realised in nature [16]. Large folding barriers might be evolutionarily preferred to prevent the native structure from partially unfolding [150]. For the protein domain BBL, however, relaxation times from various experimental measurements have been reported that cannot be reconciled with two-state (or generally multiple-state) folding and demand one-state folding with a single free energy minimum [45]. Other authors argue in favour of barrier-separated two-state folding for BBL [44].

Very recently, the Fersht group succeeded in directly observing two states for BBL by using single-molecule fluorescence resonance energy transfer (FRET) at very high time resolution ($50\,\mu$s compared to the previously possible 1 ms) [46]. Instead of indirect measurements involving the distribution of folding rates they followed a single protein and found two different states characterised by different end-to-end distances.

The structure discussed here (see Fig. 3.23) is the same as that of Ref. [46] (PDB id. 1w4h). In the simulations I protocolled a quantity comparable to the end-to-end distance from Ref. [46] but, since the unstructured N-terminal is very floppy and in the case of simulations there are no experimental limitations on protein engineering, I used the distance between the C_α-atom with residue id. 131 (corresponding to C_α-atom with index 6 within the domain) and the C_α-atom at the C-terminal end (residue id. 170, internal index 45).

In the EC-model I found a very weak but existent free energy barrier when using the distance described above, d_{ee}, as an order parameter. At low temperatures ($T = 0.07\,\epsilon$, Fig. 3.24 **(a)**) there was a single free energy minimum corresponding to the native state but at intermediate temperatures ($T = 0.08\,\epsilon$, Fig. 3.24 **(b)**) two distinct minima shortly coexisted before they merged into a single broader minimum at ($T = 0.09\,\epsilon$, Fig. 3.24 **(c)**). The authors of Ref. [46] did not report the exact values of their end-to-end distances (which were measured between different residues anyway) but found that the denatured state had lower FRET efficiency compared to the native state, corresponding to smaller distances in the unfolded state. This is also the case in my simulations. The target conformation is indicated by little red arrows in Fig. 3.24 but falls into neither of the two free energy minima. It was therefore necessary to confirm that the minimum at higher d_{ee} contains the native state as defined by contact overlap $q = 100\%$, which indeed is the case.

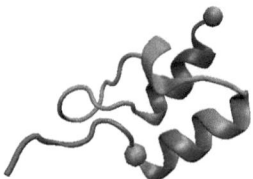

Figure 3.23: Target structure of BBL consisting of three helices. The colour scheme runs from red at the N-terminus to blue at the C-terminus. The two cyan spheres denote the C_α-atoms between which the end-to-end distance d_{ee} is measured (residue ids. 131 and 170 or internal index 6 and 45).

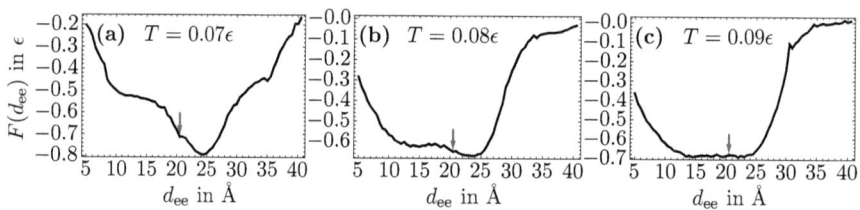

Figure 3.24: Free energy profile for BBL in the EC model. Temperature from left to right is **(a)** $T = 0.07\epsilon$, **(b)** $T = 0.08\epsilon$ and **(c)** $T = 0.09\epsilon$. The order parameter is the end-to-end distance d_{ee}, with the target conformation's end-to-end distance indicated by an arrow at $d_{ee} = 20.5$.

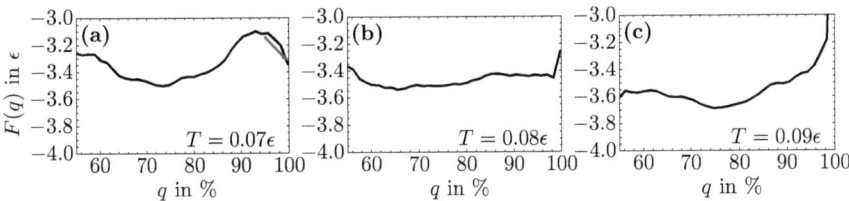

Figure 3.25: Free energy profile for BBL in the EC-model. Temperature from left to right is **(a)** $T = 0.07\epsilon$, **(b)** $T = 0.08\epsilon$ and **(c)** $T = 0.09\epsilon$. The order parameter is contact overlap q, with the native state $q = 100\%$ indicated by an arrow.

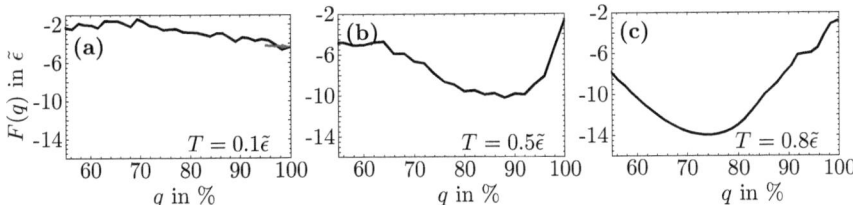

Figure 3.26: Free energy profile for BBL in the Gō-model. Temperature from left to right is **(a)** $T = 0.2\tilde{\epsilon}$, **(b)** $T = 0.5\tilde{\epsilon}$ and **(c)** $T = 0.8\tilde{\epsilon}$. The order parameter is contact overlap q, with the native state $q = 100\%$ indicated by an arrow.

Gō-model simulations of this particular protein domain have been studied before and either argued in favour of barrierless folding [151, 152] or observed that additional non-pairwise, i.e. cooperative, interactions were necessary to produce a free energy barrier [153].

I repeated these simulations using a different order parameter, namely the contact overlap q, which is a good measure of nativeness. Again, in the EC-model a barrier exists for some temperatures. This time, at low temperatures ($T = 0.07\epsilon$, Fig. 3.25) a barrier can be found but flattens for higher temperatures ($T = 0.08\epsilon$ and $T = 0.09\epsilon$, Fig. 3.25 **(b)** and **(c)**). At temperature $T = 0.09\epsilon$ the free energy minimum has moved far away from the native state. In the Gō-model only a single free energy minimum exists at all temperatures which moves away from the native state for increasing temperature (see Fig. 3.26). The EC-model can again explain the folding behaviour found in the most recent experiments but the small size and temperature-dependance of the barrier also reflect the seemingly one-state folding seen in other experiments. The Gō-model shows its now familiar behaviour of a single free energy minimum moving towards non-native conformations at increasing temperatures and cannot be reconciled with the direct observation of two distinct states [46].

Folding of the WW Domain

Both the above structures were helical, the last example protein (domain) is the WW domain consisting of a three-stranded β-sheet (Fig. 3.27) which, because of its small size and simplicity, is often investigated as a model system of β-topology [154]. Actually, the WW domain is not a single structure but a family of homologues with very similar structures but more diverse sequences. The name WW domain refers to the two conserved tryptophan residues (one-letter code W). Here, I use the structure with PDB id. 1ywj, the WW domain from formin binding protein in humans, which contains only 28 amino acids. According to the native-centric approach employed in the EC-model the sequence information is only secondary to the structure information in determining folding dynamics but this specific choice is also smaller than most other WW domain structures (which have around 35 amino acids) and lacks a rather unstructured tail.

The WW domain is abundant in eukaryotic cells and involved in signaling pathways but is also connected to diseases [155] and has been shown to form amyloids in experiment [156] and simulation [157]. Here however, I do not study interactions between several chains but only a single connected structure. Experimental and simulational observations for a single WW domain include two-state folding [154, 158], possibly even a third state has been experimentally observed for one member of the structure family [159], a denatured state consisting mostly of compact conformations [160] and helical conformations during the folding process. The latter, so far, has only been observed in MD simulations [161] but helical "overshoots" in predominantly β-domains are also reported from experiments on different proteins [162].

In order to investigate the EC-model's and the Gō-model's results on helical conformations I determined the free energy landscape in helix content and contact number as order parameters (see Fig. 3.28) using the metadynamics method. Indeed, both models see conformations at non-zero helix content although this is more pronounced in the EC-model (Fig. 3.28 (a)) where two separate minima at helix content 6 and 10 can be found. Incidentally, also the observation from all atom MD simulations that the denatured state consists mainly of compact structures is recovered in the EC-model as those minima lie at contact numbers identical to that of the native state ($N_H = 58$). In the Gō-model the sampled part of the free energy landscape reaches down to contact numbers of 20 (lower left corner) which does not agree so well with above statement on a compact denatured state but that region can hardly be regarded as a clearly defined unordered state anyway.

Two-state folding behaviour cannot be inferred or completely ruled out by either part of Fig. 3.28 as more than two free energy minima can be seen in the EC-model and one or possibly a second for the Gō-model. The grey region between a helix content of 0 and 4 is owed to the minimum size of helices and it is unclear whether it constitutes a real barrier at all or can be overcome quite easily. In this respect, helix content is not an optimal parameter to characterise distinct minima.

Nevertheless, I also looked directly at the time series of helix content in unbiased folding simulations. In the EC-model this appears to happen in two steps (Fig. 3.29). Starting from an

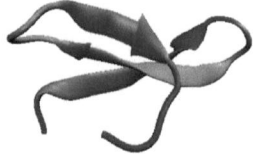

Figure 3.27: Target structure of the WW domain consisting of three β-strands. The colour scheme runs from red at the N-terminus to blue at the C-terminus.

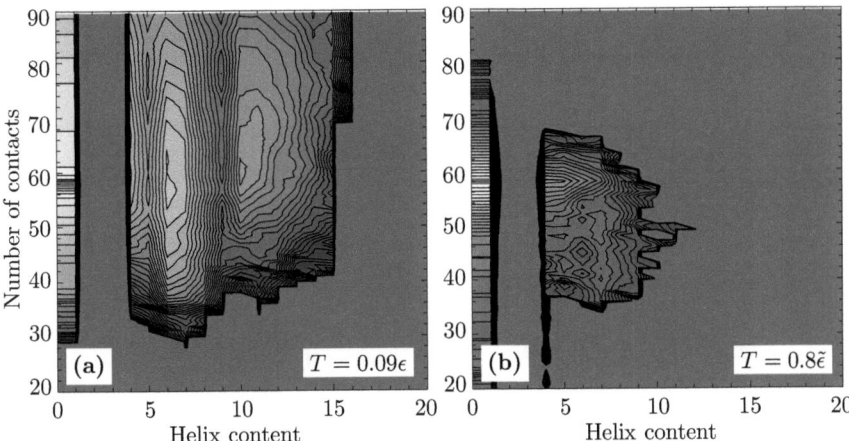

Figure 3.28: Comparison of the free energy landscape for the WW domain in the **(a)** EC- and **(b)** Gō-model. Order parameters are helix content and number of contacts. Apart from the native state minimum at zero helix content and 58 contacts, there are two additional free energy minima in the EC-model, both at higher helix content and approximately the same number of contacts. In the Gō-model, there appears one extra minimum at a helix content of 4. The unsampled region between helix content 0 and 4 is due to the minimum size of a helix element (which is 4). Colour coding and contour line spacing are the same as in Fig. 3.19 **(a)** and **(b)**, respectively.

extended chain with neither helix nor β-sheet secondary structure a helix content of 20 is quickly formed and remains stable for quite long times. There is no corresponding minimum in the free energy landscape (Fig. 3.28 **(a)**) but I hypothesise that this metastable state corresponds to the unfolded ensemble. For low temperature (Fig. 3.29 **(a)**) the system remains in that state for longer before overcoming a supposable free energy barrier (not visible in the free energy plot), for higher temperatures this happens faster (Fig. 3.29 **(b)**). Both trajectories shortly visit the free energy minimum at a helix content of 10 and the trajectory in part **(a)** even very shortly that at $N_H = 6$. The time evolution of contact number is not shown but fluctuates about the "correct" value of 58 all the time. Once the helix content has dropped to zero, the potential energy quickly follows and the native state is assumed within the free energy minimum at (0, 58). As an aside, internal energy is not able to resolve the two non-native ensembles visible in the helix content but fluctuates about the same value for both the ensemble at $N_H = 20$ and that at $N_H = 10$ although an analysis of the energy distribution might reveal a difference for these states [163]. For the Gō-model no such behaviour could be observed (data not shown) but the helix content quickly dropped to zero.

In this context, secondary structure energy E_{SS} (see Eq. (3.6)) may be useful if single conformations are to be considered. The energy smoothes conformations representing the native state as well as intermediate helical conformations. However, since these single conformations

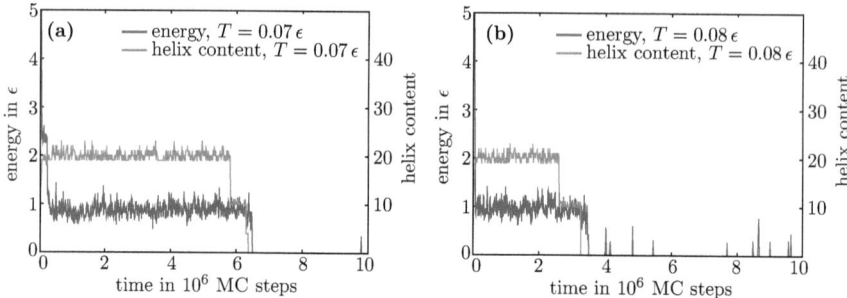

Figure 3.29: Folding time series for the WW domain in the EC-model for energy and helix content at two different temperatures, **(a)** $T = 0.07\epsilon$, **(b)** $T = 0.08\epsilon$. The energy scale is given on the left axes, the helix content scale on the right axes. A considerable amount of helical residues can be found for the all-β WW domain for quite long times. The helix content appears to drop to zero in two steps and is followed by a drop in energy. For higher temperatures the structure can escape faster from the helical trap.

are not of primary interest and including the energy term E_{SS} appears to stabilise helical conformations in folding trajectories even more, results are still shown for simulations without the additional energy.

Figures 3.30 and 3.31 show free energy landscapes in the EC-model and the Gō-model using contact overlap and RMSD as order parameters. These order parameters are strongly correlated

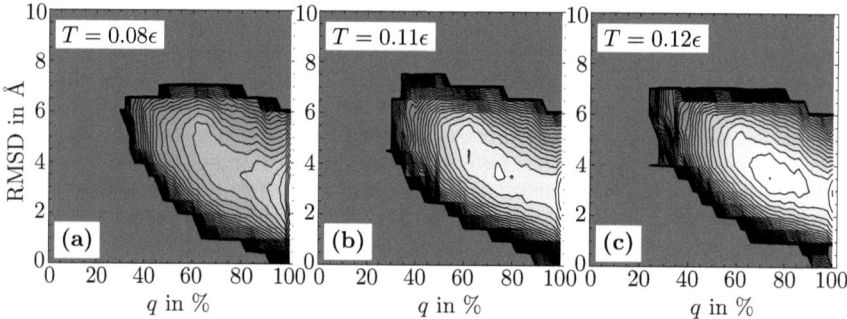

Figure 3.30: Free energy landscape for the WW domain in the EC-model. Order parameters are contact overlap q and RMSD to native state in Å. For low temperatures the free energy minimum lies close to the target state at 100 % contact overlap and an RMSD of about 3 Å. As temperature increases a new free energy minimum is formed at lower contact overlap and higher RMSD. Colour coding and contour line spacing are the same as in Fig. 3.15.

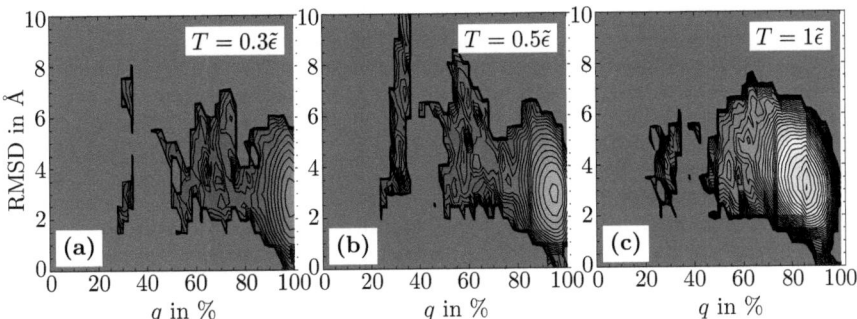

Figure 3.31: Free energy landscape for the WW domain in the Gō-model. Order parameters are contact overlap q and RMSD to native state in Å. For low temperatures the free energy minimum lies close to the target state at 100% contact overlap and an RMSD of about 3 Å. As temperature increases the free energy minimum moves to lower contact overlap and higher RMSD but no new competing minimum is formed. Colour coding and contour line spacing are the same as in Fig. 3.21.

to each other close to the native state but while the contact overlap q reflects the structure's topological similarity to the target, RMSD accounts for the geometric correctness. While the target conformation, of course, lies at an RMSD of 0 Å the native state in my model (defined by $q = 100\%$) has a higher RMSD of approximately 3 Å. Similar to the observations of the other two example proteins, in the EC-model (Fig. 3.30) growth of a new non-native minimum can be observed for increasing temperatures while in the Gō-model the existing minimum shifts towards higher values of RMSD and lower values of q. The colour scheme is the same as above with yellow introduced for the Gō-model for better contrast.

In the order parameters used here, the EC-model is again far better able to account for experimental observations or those from extensive MD simulations. All characteristic phenomena, two state-folding (with a hint at one or possibly even two additional states), a denatured state (ensemble) consisting of compact structures and helical conformations during folding, were reproduced nicely by the EC-model but not, or only less clearly, the Gō-model. Another observation that has been made in MD simulations is not visible in the present order parameters, namely that of register-shifted trap states [160]. These states have almost correctly formed β-sheets but hydrogen bonds are shifted by one residue. It is doubtful whether this would be visible in the contact pattern at all, so this is where a coarse-grained description may fail and a more detailed all-atom representation may become necessary.

3.4.6 Contact Maps as Microstates

As has been mentioned before, while low-dimensional free energy landscapes may be adequate for some models [146] it is not evident *a priori* whether all relevant properties of protein folding may be captured [164] and, in particular, whether the choice of order parameters is suitable [145]. Selection of a good order parameter may depend on the choice of model as well as the specific protein studied and it is difficult to define optimal order parameters [126].

An interesting question is whether configurations grouped together into macrostates are also microscopically similar and how structurally diverse the free energy minimum containing the native state actually is. As the model developed for this thesis allows significant distortion in space, this investigation is based on contact maps and contact overlap q used as the measure of similarity. In free energy landscapes such as Figs. 3.25 and 3.26 or Figs. 3.30 and 3.31, contact overlap to the native state has been used. This means that, in the minimum containing the native state, conformations are also very similar to each other. However, it is still not clear how structurally diverse the minimum corresponding to unfolded conformations may be. In those free energy landscapes projected onto helix content and contact number, even the native free energy minimum may hide rather diverse structures. The reason for usage of these observables in these investigations was that they are relatively easy to compare to experiments.

Therefore in a first attempt to disentangle structural complexity, the free energy minimum containing the native state in helix content and contact number parameters is examined more closely. Microstates are characterised by their contact maps instead of by full three-dimensional configurations, as those are not only more coarse-grained but have the additional advantage of being discrete. It is therefore, in principle, possible to enumerate all contact maps encountered during a simulation and count their frequencies. Attempting this for *ab initio* folding, however, is not feasible because of memory limitations. Therefore simulations were started from the native state and only its vicinity was sampled.

I ran equilibrium simulations (without metadynamics sampling) at $T = 0.08\,\epsilon$ for the villin headpiece subdomain starting from the fully folded target structure and returned contact maps every 10,000 MC steps totalling 2000 contact maps per run. From these contact frequencies and the identities of the few most frequent contact maps were extracted. Figure 3.32 **(a)** shows these contact frequencies averaged over three different runs. Darker colours denote more frequent contacts. The dark blue contacts, which have been sampled more than 5000 times, coincide with the native contacts, meaning these contacts are most common in equilibrium. Most other contacts are close to native contacts and may be formed without large rearrangements of the chain. The cluster of red and orange around indices 30 and 5 corresponds to the end terminals coming closer than in the native state and thereby forming additional contacts.

By far the most frequent single contact map was the target state's map, while most other maps have been sampled only once – again showing how vast phase space is even in this simplified representation close to the native state. In total, 978 different contact maps were encountered. Of those contact maps that were sampled more than once (ca. 5 to 20 times) most are very similar to the target contact map with a notable exception being the contact map shown in Fig. 3.32 **(b)**. This contact map was encountered 9 times, lacks all inter-helical contacts and corresponds to a three-dimensional configuration that is more extended in space than the target structure. Although, with a helix content of 29 and a contact number of 61, it is part of the native free energy minimum, its contact overlap to the target structure is only 72%.

Thus there is still structural diversity within the free energy minimum containing the native state. However, native contacts are very dominant. Helix content and contact number therefore appear to characterise the possible configurations of the villin headpiece subdomain rather faithfully. In a next step, typical configurations from the other minima discernible in the free energy landscapes might be determined although they are probably more diverse than the native state and therefore more difficult to describe.

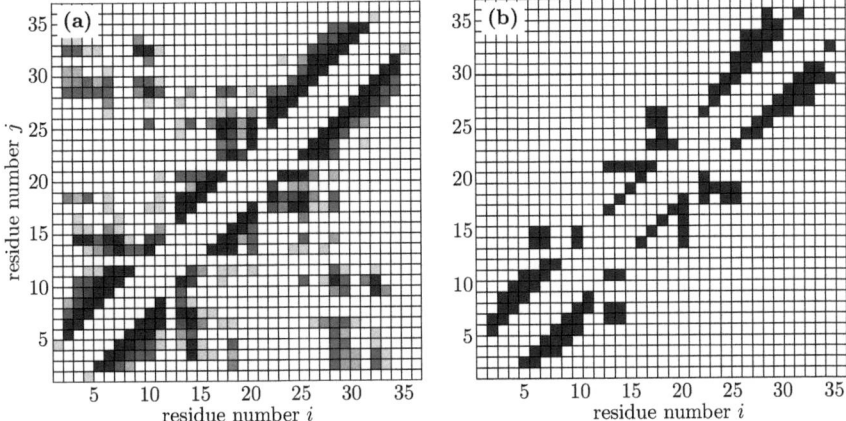

Figure 3.32: Frequent contacts and frequent contact maps: Part **(a)** shows the frequency of contacts at $T = 0.08\,\epsilon$ at equilibrium (i.e. close to the native state) for the villin headpiece in the EC-model. The colour coding is approximately logarithmic. Dark blue contacts have been sampled more than 5000 times in three accumulated runs (and happen to be the native contacts), yellow has been sampled once, orange ten times, red 100 times and purple 1000 times. Part **(b)** shows a single contact map that has been sampled nine times in a single run (of 2000 outputs). This contact map is remarkable in that it contains no inter-helical contacts.

Instead of analysing contact map distributions in free energy minima characterised by predefined order parameters, it seems more advisable to base the free energy landscape on clusters of contact maps and thus directly link the macroscopic to the microscopic picture. A new project might therefore consist in clustering contact maps by mutual contact overlaps, following the ideas of Ref. [145], or, more ambitiously, by mutual transitions in the simulation and compute free energies for the clusters. The free energy landscape could then be characterised entirely in terms of contact maps relieving the picture from the dependance on predefined order parameters. A drawback would, however, be that such clusters of contact maps are not observable in experiments making the free energy landscape obtained in simulations more difficult to link to experimental observations.

3.4.7 Heat Capacities and Folding Transitions

In all the simulations discussed above, the EC-model appeared to result in free energy landscapes of at least two different distinct macrostates while the Gō-model displayed only a single minimum which moved on the free energy landscape for changing temperatures. This difference should also be visible in plots of the specific heat of the simulations where a cooperative folding should result in a sharp peak at a well-defined transition temperature. Moreover, the number of peaks in the heat capacity curves sheds light on a possible hierarchy of distinct stages in folding.

Equilibrium simulations were run for the villin headpiece domain and heat capacities were calculated from the fluctuations in internal energy,

$$C_V = \frac{\langle U^2 \rangle - \langle U \rangle^2}{T^2}. \qquad (3.24)$$

Error bars for the heat capacity values were calculated using the jackknife method [132], possible correlation times in the equilibrium ensemble [132] were ignored. Figure 3.33 shows the heat capacity curves in (a) the EC-model and (b) the Gō-model. The most obvious observation is that the peak in the cooperative EC-model is very sharp (with its maximum at a temperature of about $T = 0.09\,\epsilon$) while the maximum in the Gō-model is very broad. The latter does not tend to zero for any temperatures simulated, so fluctuations in energy never die out entirely, which is a reliable indication of the absence of an energy gap. Upon closer examination of the heat capacity in the EC-model, there appears to be a second, less pronounced, maximum at a temperature of about $T = 0.3\,\epsilon$. The first explanation that comes to mind is that this maximum must be connected either to the formation of secondary structure elements or to the transition from extended to compact structures. I tested this hypothesis by running simulations in which the secondary structure elements were completely rigid in the conformations they have in the target structure. Only moves that changed the positions of helices relative to each other were allowed, whereas movement within helices was not. The resulting heat capacity curve shows a maximum at a similar position (see Fig. 3.33 (a) and (c), red curve) as the curve obtained without fixed secondary structure. As the model with fixed secondary structure cannot show secondary structure formation this means that the small peak is likely to correspond to the transition towards more compact structures.

In a next step, the full energy distributions for the temperatures of interest were examined instead of only the energy fluctuations (Fig. 3.34). In fact, this amounts to using E as an order parameter in free energy landscapes (only that counts instead of their logarithm are considered). The cooperative folding transition causing the sharp peak is readily visible in the energy distributions. At low temperatures a single maximum exists in the energy distribution which

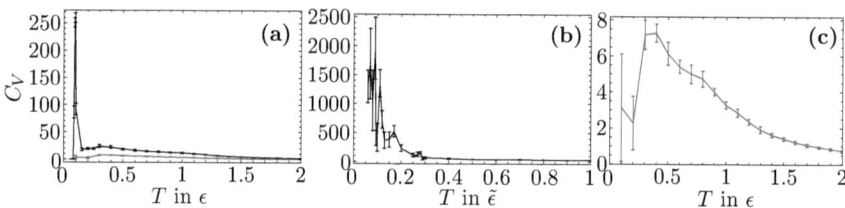

Figure 3.33: Heat capacity curves for the villin headpiece in (a) the EC-model, (b) the Gō-model and (c) the EC-model with rigid secondary structure. The red curve in (a) corresponds to (c). Note that the temperature axis for the Gō-model stops at $T = 1\,\tilde{\epsilon}$. The EC-model displays a sharp heat capacity peak at a temperature of $T \approx 0.09\,\epsilon$ and a smaller broader maximum at $T \approx 0.3\,\epsilon$. In the Gō-model heat capacity simply rises for decreasing temperature while error bars increase as well. The EC-model with fixed secondary structure also has a small rather broad peak at $T \approx 0.3\,\epsilon$.

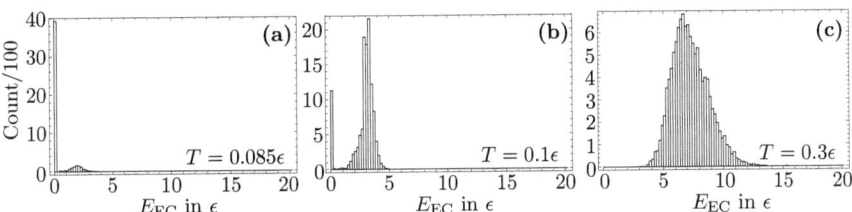

Figure 3.34: Energy distribution for the villin headpiece in the EC-model at different temperatures. **(a)** At $T = 0.085\,\epsilon$ a competing local maximum to the native energy first appears and **(b)** overtakes the first energy peak at $T = 0.1\,\epsilon$. **(c)** The second (small) peak in the heat capacity curve at $T = 0.3\,\epsilon$ is not witnessed by a bimodal energy distribution – the distribution simply broadens.

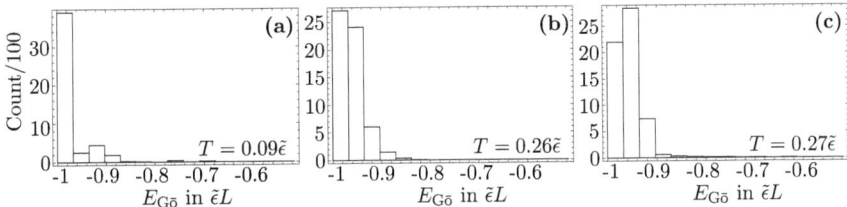

Figure 3.35: Energy distribution for the Gō-model at different temperatures. The distribution is unimodal at all times, the closest thing to a second maximum appears at $T = 0.09\,\tilde{\epsilon}$ (but note that this is only a single bin). For higher temperatures the maximum shifts to higher energies.

corresponds to proteins frozen in the native state. At $T = 0.085\,\epsilon$ higher energy structures appear at energies of approximately $E_{EC} = 2\,\epsilon$ (Fig. 3.34 **(a)**). For increasing temperatures the new maximum overtakes the native one while at the same time moving towards slightly higher energies ($T = 0.1\,\epsilon$, Fig. 3.34 **(b)**). The second maximum in heat capacity, on the other hand, is not reflected in a new maximum in energy ($T = 0.3\,\epsilon$, Fig. 3.34 **(c)**). Instead, the distribution broadens which also accounts for a rise in energy fluctuations.

Energy distributions for the Gō-model can be found in Fig. 3.35. The bin size is larger because of the discreteness of $E_{Gō}$, such that there is one possible energy value per bin. At a low temperature of $T = 0.09\,\tilde{\epsilon}$ a second maximum in the energy distribution appears very shortly but is only due to a single bin so it may well be a statistical error (Fig. 3.35 **(a)**). At increasing temperatures the maximum first broadens (Fig. 3.35 **(b)**) and then moves towards higher values of $E_{Gō}$ (Fig. 3.35 **(c)**). This behaviour is consistent with what has been observed for all the other order parameters.

In order to gain more insight about the small maximum in heat capacity in the EC-model (Fig. 3.33 **(a)** and **(c)**), the level of single conformations and contact maps was revisited. As these unfiltered structures are rather diverse, especially for high temperatures, identification of typical structures was not trivial. At the rather low temperature of $T = 0.1\,\epsilon$, which is never-

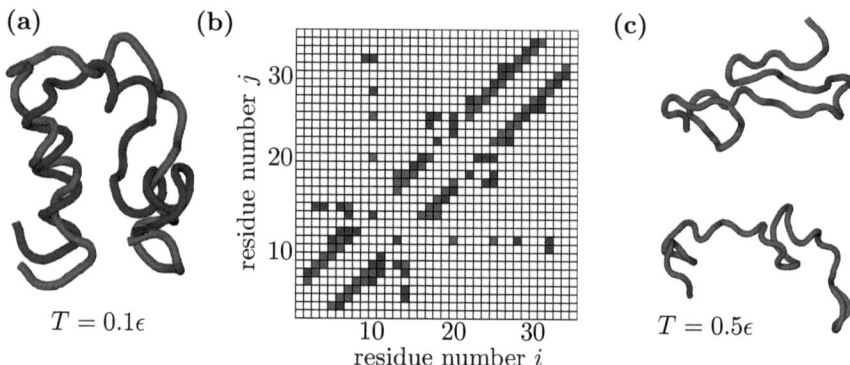

Figure 3.36: Snapshots of (villin headpiece) configurations and a contact map in the EC-model at different temperatures. Part **(a)** shows a slightly dissolved configuration at $T = 0.1\epsilon$ (red) that has been aligned to the target structure (blue) for illustration. The corresponding contact map is shown in **(b)** where contacts appearing only in the native structure are blue, contacts appearing only in the deformed conformation are red and contacts common to both structures purple. **(c)** At $T = 0.5\epsilon$ conformations no longer bear any resemblance to the target conformation. They are usually still compact but extended conformations can be found, too.

theless located at the high-temperature side of the sharp heat capacity peak, a representative structure is shown in Fig. 3.36 **(a)** with its corresponding contact map in Fig. 3.36 **(b)**. Blue indicates the target for both three-dimensional conformation and the contact map and red the example configuration, common contacts are purple in **(b)**. At $T = 0.1\,\epsilon$ typical structures still appear rather compact and with largely intact secondary structure. There are also tertiary contacts but not the same as in the native state. Beyond the second heat capacity minimum, at $T = 0.5\,\epsilon$, it is more difficult to find typical structures. In any case, helix content is not conserved anymore, instead there appear conformations carrying β-sheet-like contact patterns. As the restricted EC-model strongly encourages contact patterns reminiscent of secondary structure, these patterns still exist although they need not be the correct kind of contact pattern. At the same time, conformations become more extended so the two possible transitions (extended-compact or secondary structure formation) cannot be disentangled. However, the shift towards more extended structures, which could also be seen in the model with fixed secondary structure, seems to be necessary to dissolve and change the secondary structure elements.

3.5 Discussion

A protein model was developed based on the native state's effective connectivity (EC), a special choice of structural profile. This model is similar to Gō-models in that both models are native-centric. They rely on the knowledge of the native structure and incorporate a bias leading towards that state. They differ, however, in the important aspect that the Gō-model consists in purely additive pairwise interactions while the EC-model naturally incorporates many-body correlations.

Although folding in the EC-model is not as fast as folding in the Gō-model, the EC-model reliably folds small all-α proteins in a standard Metropolis Monte Carlo (MC) algorithm. In order to gather the necessary statistics for free energy landscapes, however, it became necessary to employ enhanced sampling methods.

Folding behaviour of three small proteins, that have been in the focus of attention of both experimentalists and theorists, was investigated and the results obtained for the EC- and the Gō-model were compared to experimental evidence and results from more detailed all-atom simulations. While for the villin headpiece subdomain the distinct intermediates observed in one Molecular Dynamics (MD) simulation [130, 131] could not be confirmed in either of the models, the EC-model was able to reproduce the existence of residual secondary structure in the denatured state as well as two-state folding behaviour that was reported from experiments. In the Gō-model folding of the villin headpiece appeared to be downhill, as for all other example proteins studied. The cooperative folding behaviour of the villin headpiece in the EC-model was also confirmed by the shape of the heat capacity curve for this model. Different folding stages are suggested based on a comparison to simulations where secondary structure elements are fixed.

For the protein domain BBL, folding behaviour is still disputed in the literature. Some experiments suggest a downhill folding scenario [45] (which is compatible with the Gō-model), whereas recently two states were directly observed in an experiment of high time-resolution single-molecule spectroscopy [46]. The EC-model shows two states for this example protein although these are only separated by a small free energy barrier. The WW domain, finally, folds undoubtedly in a two-state manner as can be reconciled with the EC-model although not with the Gō-model. Moreover, the EC-model is also in accordance with both the experimental observation that the unfolded state of the WW domain consists of compact conformations and the result from MD simulations on helical conformations in the unfolded ensemble.

The EC-model therefore extends the applicability of native-centric models beyond mere steric effects and to the description of cooperative effects. Free energy landscapes contain more detailed features in the EC-model than in the Gō-model (based on the principle of minimum frustration) which were characterised in terms of their contact topology. Using effective connectivities thus offers a new native-centric protein model that is only marginally more complicated than the Gō-model but in better agreement with results obtained from experiments and detailed MD simulations.

4 Conclusion and Outlook

It is known that one-dimensional structural profiles can be employed to describe protein structure, identify protein domains [5] or efficiently compare structures of different proteins [63]. This thesis shows that the usefulness of structural profiles extends to protein structure prediction and folding dynamics.

In the context of structure prediction, it is especially important that structure profiles can be inferred from sequences with quite good accuracy and much more easily than, for instance, contact maps [61]. In this thesis, exact and predicted ECs have been employed to select promising near-native candidates from a coarse-grained structure set. This is of practical importance, as refinement in protein structure prediction is costly and time-consuming and therefore should be limited to candidate structures that are worth the effort. Filtering by predicted ECs poses an advantage compared to established selection methods, in particular if the initial structure set is only of moderate quality or if only few structures are to be selected. Moreover, it is more versatile than the clustering approach which relies on extensive sampling of conformation space that becomes troublesome for increasing protein size.

In a next project building on these results, it might be worthwhile to integrate the structural profile into the structure generation step in order to improve the quality of the initial set of coarse-grained structures. Predicted structural profiles might be treated as a constraint in structure generation, similar to sparse experimental data from NMR chemical shifts [85, 86]. As the effective connectivity involves calculation of a matrix eigensystem it could prove advantageous to use the contact vector instead. The contact vector can be calculated faster and predicted with comparable accuracy. Its drawback, i.e. its degeneracy [64], might not be as severe if profiles are predicted and therefore already inexact and if these predicted profiles are then used as one constraint among several energy terms.

In the context of protein folding, the EC-model provides a simple yet cooperative protein model based on the native structure. The crucial difference between Gō-model and EC-model is that the EC-model incorporates non-pairwise interactions between amino acids. In the Gō-model native contacts are always favourable and non-native contacts either ignored or repulsive, whereas in the EC-model a contact between two amino acids can be either attractive or repulsive depending on how many contacts those amino acids already have. Energies are therefore context-dependent which introduces non-local interactions into the model. For the EC-model's folding behaviour this means that amino acids have to assemble into their correct structure in a cooperative fashion, as can be seen when comparing heat capacity curves. As the EC-model is a native-centric model that does not rely on the principle of minimum frustration, it may help to disentangle these two assumptions that are often made in simple protein models.

In the Gō-model the only possibility for amino acids to interact, if they are not in contact in the native state, is to do so sterically. If part of the chain blocks the way for a native contact, this part has to be moved away before the contact can be formed. Cooperativity in the Gō-model is therefore much weaker and only steric effects of folding may be captured. Conceptually, while the Gō-model strives to form all contacts present in the native state the EC-model aims at satisfying each amino acid's contact propensity which also relates to the picture of hydrophobic interactions between residues.

A way to use the Gō-model profitably is to add it to more sophisticated potentials as a bias towards the native structure and thus ensure folding while also including more specific interactions [102] or to incorporate sequence information only where necessary [165]. The advantage of the EC-model, however, is that it separates purely structural information from sequence input so the question whether the native structure largely prescribes the folding pathway may be addressed.

A possible extension of the present model is the use of noisy structural profiles. In a way, this has been investigated in the context of structure prediction where exact profiles were substituted by predicted ones. In order to analyse the effect on protein folding dynamics in a controlled way, noise could be added to the exact EC and the resulting free energy landscapes explored using the sampling techniques described. Besides, free energy landscapes could also be investigated in more detail by using clusters of contact maps as macrostates in order to become independent of predefined order parameters.

Finally, the difference between the EC-model and the Gō-model might be further examined by constructing a Gō-type energy from average energy contributions of native contacts between two amino acids in the EC-model. This would result in a Gō-model with heterogeneous energy terms that are, however, still pairwise and additive. The necessity to include higher-order correlations in order to produce two-state kinetics might thus be addressed.

The applicability of structural profiles to the area of protein folding, both in structure prediction and in folding dynamics, has thus been demonstrated and compared to established models and methods. In both fields structural profiles can be profitably employed.

A RMSD and TM-Score Distribution of Candidate Structures

This appendix summarises the RMSD and TM-score distributions for all proteins from Chapter 2 for which candidate sets were generated using the Rosetta method. Proteins are sorted by length and smaller proteins, namely those with less than 200 amino acids, show some structure in RMSD distributions (all left-hand figures from Fig. A.1 **(a)** to **(w)**). This evidences that several (free) energy minima of conformation space have been sampled by the structure generation protocol. The only exception is the minor coat protein G3P (PDB id. 1fgp) which is rather small, as it consists of only 70 amino acids. However, it displays a Gaussian distribution of RMSD values and a rather poor candidate set as measured by either RMSD or TM-score. Another protein for which the candidate set was of poor quality is a single chain of the dimeric trigger factor mutant F44L (PDB id. 1p9yA) which is also the longest among the small proteins and is difficult in that it contains two β-sheets that are distant in sequence but have to meet in the structure. Particularly good candidate sets were created for the N-terminal domain of phage 434 repressor (PDB id. 1r69) and the immunoglobulin binding domain of protein G (PDB id. 1gb1).

For all proteins the dotted black line indicates an RMSD of 3 Å from the native structure which is usually regarded as the threshold of good predictions that are likely to converge further to the native structure in refinement. The solid red line marks the mean for each distribution and the dashed red line denotes one standard deviation below this mean. Everything to the left of that dashed line is considered as relatively good throughout Chapter 2.

TM-score distributions (right-hand histograms in Fig. A.1) show less structure than RMSD distributions and are usually unimodal but otherwise agree with RMSD distributions in which candidate sets are of high or low quality. A structure is considered truly near-native if TM-score is below 0.6 (hence the dotted black line at $1-\text{TM-score}=0.4$) but, again, relative candidate quality is assessed by whether or not the candidate has a TM-score above the mean plus one standard deviation.

For proteins longer than 200 amino acids (Fig. A.1 from **(y)** onward) the Gaussian shape of RMSD distributions is typical. This means that no significant relation between candidates and target structure is detectable, as a Gaussian distribution is what is expected for the RMSD between two unrelated but compact structures. The only exception where a protein deviates from this behaviour is the Dbl homology domain from beta-PIX (PDB id. 1by1) where two maxima exist in the distribution. The candidate sets hardly contained any structures of TM-score above 0.5 or RMSD below 7 Å (see Fig. A.1 from **(y)** onward). The Gaussian curves plotted in the RMSD histograms for larger proteins are fitted such that mean and variance agree with values from the candidate distribution. Those approximations were used to estimate the number of structures which would be necessary in order to expect one structure of RMSD below

2imf	1volA	1ix9A	1f5x	1gk9A	1by1
$8.8 \cdot 10^{11}$	$2.4 \cdot 10^4$	$1.1 \cdot 10^{12}$	$1.2 \cdot 10^{13}$	$2.0 \cdot 10^8$	$3.4 \cdot 10^3$

Table A.1: Estimated number of candidates to expect one structure below RMSD=5 Å. In fact, 10,000 candidates were generated, the only protein for which this should have been sufficient is 1by1.

5 Å. The estimates are given in Table A.1 and are realistic for all proteins except 1by1 where the number of low-RMSD candidates is over-estimated and the number of necessary candidates thus under-estimated because of the two maxima in the real distribution. If only the first maximum in the distribution for the protein with PDB id. 1by1 (at lower RMSDs) is fitted, the number of necessary structures is of the order of 10^4 to expect one good structure.

According to these estimates a low-RMSD decoy (below 5 Å) should have been found in the candidate sets of 10,000 structures only for the protein 1by1 and possibly chain A of the transcription factor IIB (PDB id. 1volA) but is beyond reach for all other target proteins. Actually, such a structure below an RMSD of 5 Å was found for none of the longer proteins. However, instead of increasing the size of the candidate sets by orders of magnitude, longer structures were obtained from the 8th Critical Assessment of Techniques for Protein Structure Prediction (CASP8) [87], see Chapter 2.

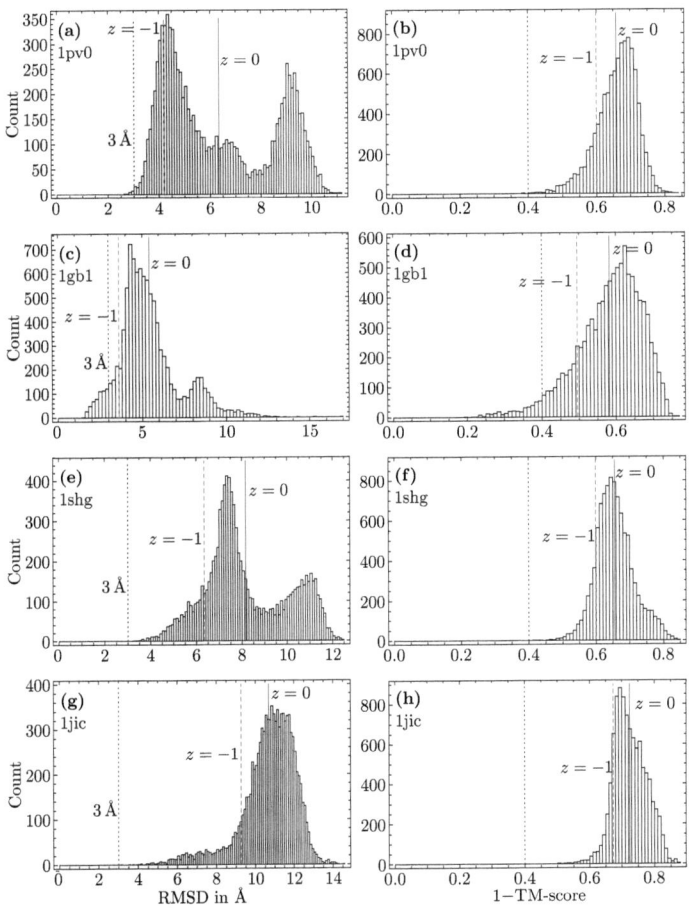

Figure A.1: Distribution of RMSD and $1-$TM-score for proteins 1pv0, 1gb1, 1shg and 1jic

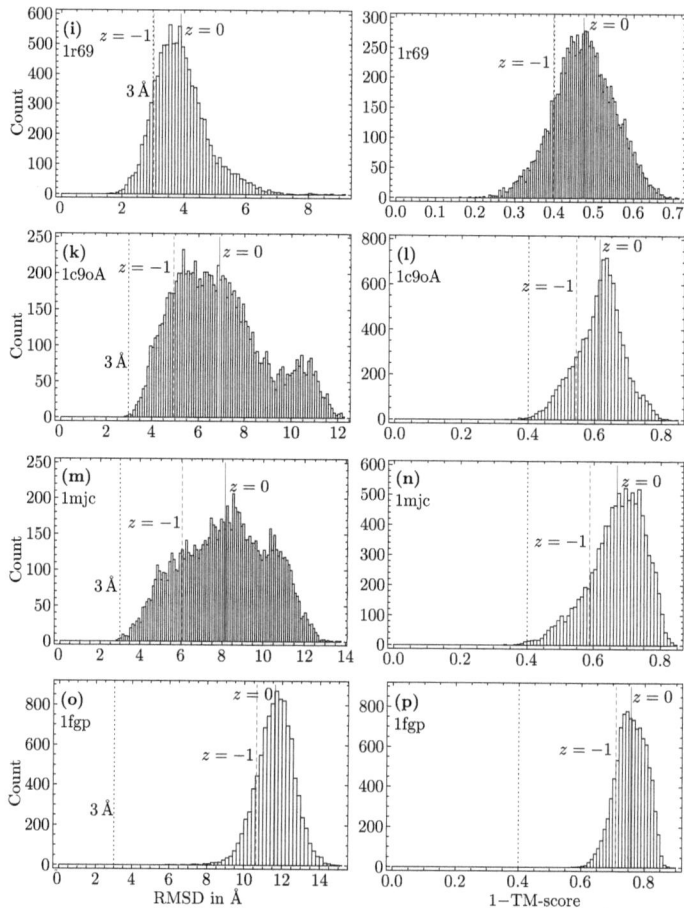

Figure A.1: Distribution of RMSD and 1−TM-score for proteins 1r69, 1c9oA, 1mjc and 1fgp

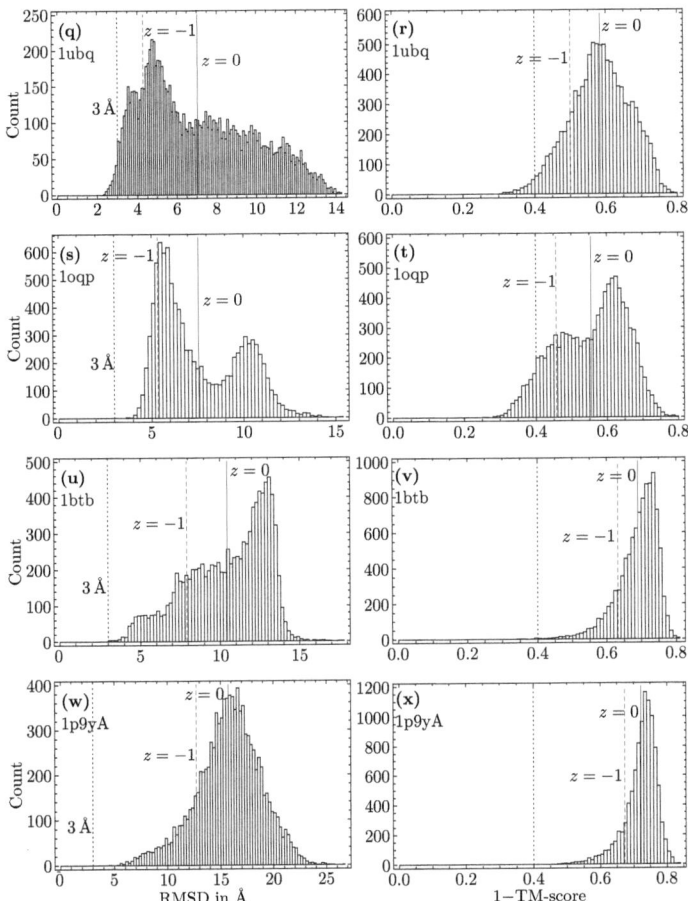

Figure A.1: Distribution of RMSD and 1−TM-score for proteins 1ubq, 1oqp, 1btb and 1p9yA

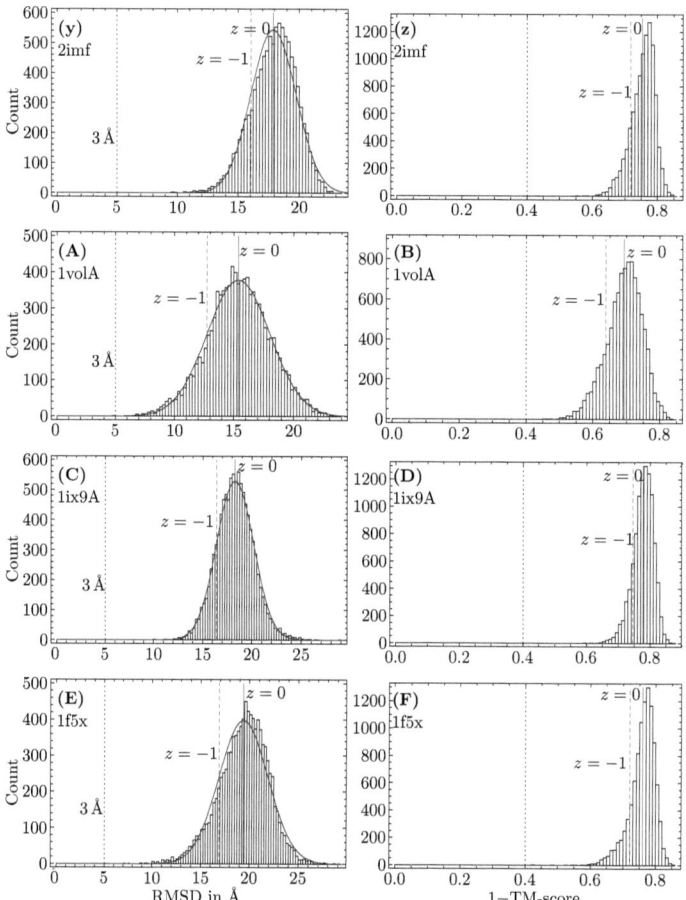

Figure A.1: Distribution of RMSD and 1−TMscore for proteins 2imf, 1volA, 1ix9A and 1f5x. The RMSD distribution of these longer proteins approaches a Gaussian curve.

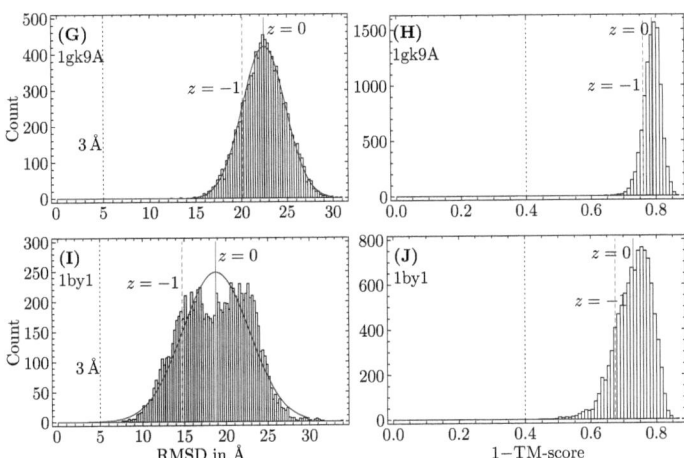

Figure A.1: Distribution of RMSD and $1-$TMscore for proteins 1gk9A and 1by1. The RMSD histogram of 1by1 does not acquire the Gaussian shape typical for longer proteins.

Acknowledgements

I gratefully acknowledge excellent advice by Prof. Markus Porto including perfect availability and encouragement. I thank Dr Michele Vendruscolo for his kind hospitality in Cambridge, inspiring discussions and for sharing his ideas, Dr Andrea Cavalli for help with the early stages of the structure prediction project and Jonas Minning for providing predicted structural profiles. Many thanks go to my parents for support and encouragement throughout the years. Finally, I want to thank Anne, Christiane and Johannes for proof-reading this thesis from a biologist's, chemist's and physicist's point of view.

Bibliography

[1] C. M. Dobson. Protein folding and misfolding, *Nature*, **426**, 884–890, (2003).

[2] C. B. Anfinsen. Principles that govern the folding of protein chains, *Science*, **181**, 223–230, (1973).

[3] D. Vitkup, E. Melamud, J. Moult, and C. Sander. Completeness in structural genomics, *Nat. Struct. Mol. Biol.*, **8**, 559–566, (2001).

[4] N. O'Toole, M. Grabowski, Z. Otwinowski, W. Minor, and M. Cygler. The structural genomics experimental pipeline: Insights from global target lists, *Proteins*, **56**, 201–210, (2004).

[5] L. Holm and C. Sander. Parser for protein folding units, *Proteins*, **19**, 256–268, (1994).

[6] U. Bastolla, A. R. Ortiz, M. Porto, and F. Teichert. Effective connectivity profile: A structural representation that evidences the relationship netween protein structures and sequences, *Proteins*, **73**, 872–888, (2008).

[7] U. Bastolla, M. Porto, H. E. Roman, and M. Vendruscolo. Principal eigenvector of contact matrices and hydrophobicity profiles in proteins, *Proteins*, **58**, 22–30, (2005).

[8] D. Chivian and D. Baker. Homology modeling using parametric alignment ensemble generation with consensus and energy-based model selection, *Nucl. Acids Res.*, **34**, e112, (2006).

[9] R. Bonneau and D. Baker. Ab initio protein structure prediction: progress and prospects, *Annu. Rev. Biophy. Biom.*, **30**, 173–189, (2001).

[10] C. A. Rohl, C. E. Strauss, K. M. Misura, and D. Baker. Protein structure prediction using Rosetta, *Methods Enzymol.*, **383**, 66–93, (2004).

[11] M. R. Betancourt and J. Skolnick. Finding the needle in a haystack: educing native folds from ambiguous *ab initio* protein structure predictions, *J. Comput. Chem.*, **22**, 339–353, (2001).

[12] H. Taketomi, Y. Ueda, and N. Gō. Studies on protein folding, unfolding and fluctuations by computer simulation, *Int. J. Pept. Prot. Res.*, **7**, 445–459, (1975).

[13] N. D. Socci, J. N. Onuchic, and P. G. Wolynes. Diffusive dynamics of the reaction coordinate for protein folding funnels, *J. Chem. Phys.*, **104**, 5860–5868, (1996).

[14] J. D. Honeycutt and D. Thirumalai. The nature of folded states of globular proteins, *Biopolymers*, **32**, 695–709, (1992).

[15] S. Schnabel, M. Bachmann, and W. Janke. Two-State folding, folding through intermediates, and metastability in a minimalistic Hydrophobic-Polar model for proteins, *Phys. Rev. Lett.*, **98**, 048103–4, (2007).

[16] J. D. Bryngelson, J. N. Onuchic, N. D. Socci, and P. G. Wolynes. Funnels, pathways, and the energy landscape of protein folding: A synthesis, *Proteins*, **21**, 167–195, (1995).

[17] B. Roux and T. Simonson. Implicit solvent models, *Biophys. Chem.*, **78**, 1–20, (1999).

[18] R. Zhou. Free energy landscape of protein folding in water: Explicit vs. implicit solvent, *Proteins*, **53**, 148–161, (2003).

[19] J. W. Ponder and D. A. Case. Force fields for protein simulations, *Adv. Protein Chem.*, **66**, 27–85, (2003).

[20] J. Lee, A. Liwo, and H. A. Scheraga. Energy-based *de novo* protein folding by conformational space annealing and an off-lattice united-residue force field, *Proc. Natl. Acad. Sci. U.S.A.*, **96**, 2025–2030, (1999).

[21] A. Korkut and W. A. Hendrickson. A force field for virtual atom molecular mechanics of proteins, *Proc. Natl. Acad. Sci. U.S.A.*, **106**, 15667–15672, (2009).

[22] M. Cieplak, T. X. Hoang, and M. S. Li. Scaling of folding properties in simple models of proteins, *Phys. Rev. Lett.*, **83**, 1684, (1999).

[23] Y. A. Arnautova and H. A. Scheraga. Use of decoys to optimize an all-atom force field including hydration, *Biophys. J.*, **95**, 2434–2449, (2008).

[24] V. Muñoz, E. R. Henry, J. Hofrichter, and W. A. Eaton. A statistical mechanical model for β-hairpin kinetics, *Proc. Natl. Acad. Sci. U.S.A.*, **95**, 5872–5879, (1998).

[25] V. Muñoz and W. A. Eaton. A simple model for calculating the kinetics of protein folding from three-dimensional structures, *Proc. Natl. Acad. Sci. U.S.A.*, **96**, 11311–11316, (1999).

[26] J. D. Bryngelson and P. G. Wolynes. Spin glasses and the statistical mechanics of protein folding, *Proc. Natl. Acad. Sci. U.S.A.*, **84**, 7524–7528, (1987).

[27] D. J. Wales, *Energy Landscapes*. Cambridge: Cambridge University Press, (2003).

[28] W. Hendrickson. Determination of macromolecular structures from anomalous diffraction of synchrotron radiation, *Science*, **254**, 51–58, (1991).

[29] H. Hansen-Goos, R. Roth, K. Mecke, and S. Dietrich. Solvation of proteins: Linking thermodynamics to geometry, *Phys. Rev. Lett.*, **99**, 128101–4, (2007).

[30] G. N. Ramachandran, C. Ramakrishnan, and V. Sasisekharan. Stereochemistry of polypeptide chain configurations, *J. Mol. Biol.*, **7**, 95–99, (1963).

[31] L. Pauling, R. B. Corey, and H. R. Branson. The structure of proteins: Two hydrogen-bonded helical configurations of the polypeptide chain, *Proc. Natl. Acad. Sci. U.S.A.*, **37**, 205–211, (1951).

[32] R. Breslow and Z. Cheng. On the origin of terrestrial homochirality for nucleosides and amino acids, *Proc. Natl. Acad. Sci. U.S.A.*, **106**, 9144–9146, (2009).

[33] A. Ryle, F. Sanger, L. Smith, and R. Kitai. Disulphide bonds of insulin, *Biochem. J.*, **60**, 541–556, (1955).

[34] J. C. Kendrew, G. Bodo, H. M. Dintzis, R. G. Parrish, H. Wyckoff, and D. C. Phillips. A three-dimensional model of the myoglobin molecule obtained by x-ray analysis, *Nature*, **181**, 662–666, (1958).

[35] PDB website http://www.rcsb.org/pdb/home/home.do.

[36] A. E. Franke, D. E. Danley, F. S. Kaczmarek, S. J. Hawrylik, R. D. Gerard, S. Lee, and K. F. Geoghegan. Expression of human plasminogen activator inhibitor type-1 (PAI-1) in Escherichia coli as a soluble protein comprised of active and latent forms. Isolation and crystallization of latent PAI-1, *Biochim. Biophys. Acta*, **1037**, 16–23, (1990).

[37] D. Baker and D. A. Agard. Kinetics versus thermodynamics in protein folding, *Biochemistry*, **33**, 7505–7509, (1994).

[38] S. S. Jaswal, J. L. Sohl, J. H. Davis, and D. A. Agard. Energetic landscape of α-lytic protease optimizes longevity through kinetic stability, *Nature*, **415**, 343–346, (2002).

[39] J. Mittal and R. B. Best. Thermodynamics and kinetics of protein folding under confinement, *Proc. Natl. Acad. Sci. U.S.A.*, **105**, 20233–20238, (2008).

[40] R. J. Ellis and A. P. Minton. Cell biology: Join the crowd, *Nature*, **425**, 27–28, (2003).

[41] C. Levinthal. Are there pathways for protein folding?, *J. Chim. Phys.*, **65**, 44, (1968).

[42] R. Zwanzig, A. Szabo, and B. Bagchi. Levinthal's paradox, *Proc. Natl. Acad. Sci. U.S.A.*, **89**, 20–22, (1992).

[43] K. A. Dill and H. S. Chan. From Levinthal to pathways to funnels, *Nat. Struct. Mol. Biol.*, **4**, 10–19, (1997).

[44] N. Ferguson, P. J. Schartau, T. D. Sharpe, S. Sato, and A. R. Fersht. One-state downhill versus conventional protein folding, *J. Mol. Biol.*, **344**, 295–301, (2004).

[45] P. Li, F. Y. Oliva, A. N. Naganathan, and V. Muñoz. Dynamics of one-state downhill protein folding, *Proc. Natl. Acad. Sci. U.S.A.*, **106**, 103–108, (2009).

[46] F. Huang, L. Ying, and A. R. Fersht. Direct observation of barrier-limited folding of BBL by single-molecule fluorescence resonance energy transfer, *Proc. Natl. Acad. Sci. U.S.A.*, **106**, 16239–16244, (2009).

[47] A. G. Murzin, S. E. Brenner, T. Hubbard, and C. Chothia. SCOP: a structural classification of proteins database for the investigation of sequences and structures., *J. Mol. Biol.*, **247**, 536–540, (1995).

[48] C. A. Orengo, A. D. Michie, S. Jones, D. T. Jones, M. B. Swindells, and J. M. Thornton. CATH – a hierarchic classification of protein domain structures, *Structure*, **5**, 1093–1108, (1997).

[49] M. Levitt. Nature of the protein universe, *Proc. Natl. Acad. Sci. U.S.A.*, **106**, 11079–11084, (2009).

[50] C. Chothia and A. Lesk. The relation between the divergence of sequence and structure in proteins., *EMBO J.*, **5**, 826, 823, (1986).

[51] P. A. Alexander, Y. He, Y. Chen, J. Orban, and P. N. Bryan. The design and characterization of two proteins with 88% sequence identity but different structure and function, *Proc. Natl. Acad. Sci. U.S.A.*, **104**, 11963–11968, (2007).

[52] C. G. Roessler, B. M. Hall, W. J. Anderson, W. M. Ingram, S. A. Roberts, W. R. Montfort, and M. H. J. Cordes. Transitive homology-guided structural studies lead to discovery of cro proteins with 40% sequence identity but different folds, *Proc. Natl. Acad. Sci. U.S.A.*, **105**, 2343–2348, (2008).

[53] B. R. Brooks, R. E. Bruccoleri, B. D. Olafson, D. J. States, S. Swaminathan, and M. Karplus. CHARMM: a program for macromolecular energy, minimization, and dynamics calculations, *J. Comput. Chem.*, **4**, 187–217, (1983).

[54] W. D. Cornell, P. Cieplak, C. I. Bayly, I. R. Gould, K. M. Merz, D. M. Ferguson, D. C. Spellmeyer, T. Fox, J. W. Caldwell, and P. A. Kollman. A second generation force field for the simulation of proteins, nucleic acids, and organic molecules, *J. Am. Chem. Soc.*, **117**, 5179–5197, (1995).

[55] D. V. D. Spoel, E. Lindahl, B. Hess, G. Groenhof, A. E. Mark, and H. J. C. Berendsen. GROMACS: fast, flexible, and free, *J. Comput. Chem.*, **26**, 1701–1718, (2005).

[56] A. A. Canutescu, A. A. Shelenkov, and R. L. Dunbrack. A graph-theory algorithm for rapid protein side-chain prediction, *Protein Sci.*, **12**, 2001–2014, (2003).

[57] A. P. Heath, L. E. Kavraki, and C. Clementi. From coarse-grain to all-atom: Toward multiscale analysis of protein landscapes, *Proteins*, **68**, 646–661, (2007).

[58] Q. Yang and S. Sze. Predicting protein folding pathways at the mesoscopic level based on native interactions between secondary structure elements, *BMC Bioinformatics*, **9**, 320, (2008).

[59] M. Vendruscolo, E. Kussel, and E. Domany. Recovery of protein structure from contact maps, *Fold. Des.*, **2**, 295–306, (1997).

[60] M. Vendruscolo, R. Najmanovich, and E. Domany. Protein folding in contact map space, *Phys. Rev. Lett.*, **82**, 656, (1999).

[61] A. Vullo, I. Walsh, and G. Pollastri. A two-stage approach for improved prediction of residue contact maps, *BMC Bioinformatics*, **7**, 180, (2006).

[62] M. Vassura, L. Margara, P. D. Lena, F. Medri, P. Fariselli, and R. Casadio. FT-COMAR: fault tolerant three-dimensional structure reconstruction from protein contact maps, *Bioinformatics*, **24**, 1313–1315, (2008).

[63] F. Teichert, U. Bastolla, and M. Porto. SABERTOOTH: protein structural alignment based on a vectorial structure representation, *BMC Bioinformatics*, **8**, 425, (2007).

[64] A. Kabakçioğlu, I. Kanter, M. Vendruscolo, and E. Domany. Statistical properties of contact vectors, *Phys. Rev. E*, **65**, 041904, (2002).

[65] M. Porto, U. Bastolla, H. E. Roman, and M. Vendruscolo. Reconstruction of protein structures from a vectorial representation, *Phys. Rev. Lett.*, **92**, 218101, (2004).

[66] M. Martí-Renom, A. Stuart, A. Fiser, R. Sánchez, F. Melo, and A. Sali. Comparative protein structure modeling of genes and genomes, *Annu. Rev. Bioph. Biom.*, **29**, 325, 291, (2000).

[67] B. Al-Lazikani. Protein structure prediction, *Curr. Opin. Chem. Biol.*, **5**, 56, 51, (2001).

[68] R. Jauch, H. C. Yeo, P. R. Kolatkar, and N. D. Clarke. Assessment of CASP7 structure predictions for template free targets, *Proteins*, **69**, 57–67, (2007).

[69] J. Kopp, L. Bordoli, J. N. Battey, F. Kiefer, and T. Schwede. Assessment of CASP7 predictions for template-based modeling targets, *Proteins*, **69**, 38–56, (2007).

[70] A. Kryshtafovych, Česlovas. Venclovas, K. Fidelis, and J. Moult. Progress over the first decade of CASP experiments, *Proteins*, **61**, 225–236, (2005).

[71] A. Koliński and J. M. Bujnicki. Generalized protein structure prediction based on combination of fold-recognition with *de novo* folding and evaluation of models, *Proteins*, **61 S7**, 84–90, (2005).

[72] P. Bradley, K. M. S. Misura, and D. Baker. Toward high-resolution *de novo* structure prediction for small proteins, *Science*, **309**, 1868–1871, (2005).

[73] A. Kryshtafovych, K. Fidelis, and J. Moult. Progress from CASP6 to CASP7, *Proteins*, **69**, 194–207, (2007).

[74] Y. Zhang. Template-based modeling and free modeling by I-TASSER in CASP7, *Proteins*, **69**, 108–117, (2007).

[75] O. Schueler-Furman, C. Wang, P. Bradley, K. Misura, and D. Baker. Progress in modeling of protein structures and interactions, *Science*, **310**, 638–642, (2005).

[76] K. Wolff, M. Vendruscolo, and M. Porto. Efficient identification of near-native conformations in *ab initio* protein structure prediction using structural profiles, *Proteins*, **78**, 249–258, (2010).

[77] K. M. S. Misura and D. Baker. Progress and challenges in high-resolution refinement of protein structure models, *Proteins*, **59**, 15–29, (2005).

[78] C. M. Summa and M. Levitt. Near-native structure refinement using *in vacuo* energy minimization, *Proc. Natl. Acad. Sci. U.S.A.*, **104**, 3177–3182, (2007).

[79] A. Jagielska, L. Wroblewska, and J. Skolnick. Protein model refinement using an optimized physics-based all-atom force field, *Proc. Natl. Acad. Sci. U.S.A.*, **105**, 8268–8273, (2008).

[80] H. Gong, P. J. Fleming, and G. D. Rose. Building native protein conformation from highly approximate backbone torsion angles, *Proc. Natl. Acad. Sci. U.S.A.*, **102**, 16227–16232, (2005).

[81] A. W. Stumpff-Kane and M. Feig. A correlation-based method for the enhancement of scoring functions on funnel-shaped energy landscapes, *Proteins*, **63**, 155–164, (2006).

[82] J. Qiu, W. Sheffler, D. Baker, and W. S. Noble. Ranking predicted protein structures with support vector regression, *Proteins*, **71**, 1175–82, (2008).

[83] J. M. Bujnicki, A. Elofsson, D. Fischer, and L. Rychlewski. Structure prediction meta server, *Bioinformatics*, **17**, 750–751, (2001).

[84] J. Kosiński, M. J. Gajda, I. A. Cymerman, M. A. Kurowski, M. Pawlowski, M. Boniecki, A. Obarska, G. Papaj, P. Sroczynska-Obuchowicz, K. L. Tkaczuk, P. Sniezynska, J. M. Sasin, A. Augustyn, J. M. Bujnicki, and M. Feder. FRankenstein becomes a cyborg: The automatic recombination and realignment of fold recognition models in CASP6, *Proteins*, **61**, 106–113, (2005).

[85] A. Cavalli, X. Salvatella, C. M. Dobson, and M. Vendruscolo. Protein structure determination from NMR chemical shifts, *Proc. Natl. Acad. Sci. U.S.A.*, **104**, 9615–9620, (2007).

[86] Y. Shen, O. Lange, F. Delaglio, P. Rossi, J. M. Aramini, G. Liu, A. Eletsky, Y. Wu, K. K. Singarapu, A. Lemak, A. Ignatchenko, C. H. Arrowsmith, T. Szyperski, G. T. Montelione, D. Baker, and A. Bax. Consistent blind protein structure generation from NMR chemical shift data, *Proc. Natl. Acad. Sci. U.S.A.*, **105**, 4685–4690, (2008).

[87] CASP8 website http://predictioncenter.org/download_area/casp8/server_predictions as of June 15, 2009.

[88] S. Altschul, T. Madden, A. Schaffer, J. Zhang, Z. Zhang, W. Miller, and D. Lipman. Gapped BLAST and PSI-BLAST: a new generation of protein database search programs, *Nucl. Acids Res.*, **25**, 3389–3402, (1997).

[89] D. T. Jones. Protein secondary structure prediction based on position-specific scoring matrices, *J. Mol. Biol.*, **292**, 195–202, (1999).

[90] J. Minning, *Proteinstrukturvergleich und Proteinsequenz/-struktur Zuordnung*. Diplomarbeit, (2008).

[91] Y. Zhang and J. Skolnick. Scoring function for automated assessment of protein structure template quality, *Proteins*, **57**, 702–710, (2004).

[92] W. Kabsch. A solution for the best rotation to relate two sets of vectors, *Acta Crystallogr.*, **A32**, 922–923, (1976).

[93] W. Kabsch. A discussion of the solution for the best rotation to relate two sets of vectors, *Acta Crystallogr.*, **A34**, 827–828, (1978).

[94] E. Coutsias, C. Seok, and K. Dill. Using quaternions to calculate RMSD, *J. Comput. Chem.*, **25**, 1857, 1849, (2004).

[95] S. Wallin, J. Farwer, and U. Bastolla. Testing similarity measures with continuous and discrete protein models, *Proteins*, **50**, 50—144, (2002).

[96] Rosetta forum http://boinc.bakerlab.org/rosetta/forum_thread.php?id=1453#14335 as of June 15, 2009.

[97] C. Clementi, H. Nymeyer, and J. N. Onuchic. Topological and energetic factors: What determines the structural details of the transition state ensemble and en-route intermediates for protein folding? An investigation for small globular proteins, *J. Mol. Biol.*, **298**, 937–953, (2000).

[98] N. Koga and S. Takada. Roles of native topology and chain-length scaling in protein folding: A simulation study with a Gō-like model, *J. Mol. Biol.*, **313**, 171–180, (2001).

[99] B. T. Andrews, S. Gosavi, J. M. Finke, J. N. Onuchic, and P. A. Jennings. The dual-basin landscape in GFP folding, *Proc. Natl. Acad. Sci. U.S.A.*, **105**, 12283–12288, (2008).

[100] C. Clementi. Coarse-grained models of protein folding: toy models or predictive tools?, *Curr. Opin. Struct. Biol.*, **18**, 10–15, (2008).

[101] H. Kaya and H. S. Chan. Solvation effects and driving forces for protein thermodynamic and kinetic cooperativity: How adequate is native-centric topological modeling?, *J. Mol. Biol.*, **326**, 911–931, (2003).

[102] A. Kleiner and E. Shakhnovich. The mechanical unfolding of ubiquitin through all-atom Monte Carlo simulation with a Gō-type potential, *Biophys. J.*, **92**, 2054–2061, (2007).

[103] D. E. Makarov and K. W. Plaxco. The topomer search model: A simple, quantitative theory of two-state protein folding kinetics, *Protein Sci.*, **12**, 17–26, (2003).

[104] A. Y. Istomin, D. J. Jacobs, and D. R. Livesay. On the role of structural class of a protein with two-state folding kinetics in determining correlations between its size, topology, and folding rate, *Protein Sci.*, **16**, 2564–2569, (2007).

[105] M. M. Gromiha, A. M. Thangakani, and S. Selvaraj. FOLD-RATE: prediction of protein folding rates from amino acid sequence, *Nucl. Acids Res.*, **34**, W70–74, (2006).

[106] T. X. Hoang, A. Trovato, F. Seno, J. R. Banavar, and A. Maritan. Geometry and symmetry presculpt the free-energy landscape of proteins, *Proc. Natl. Acad. Sci. USA*, **101**, 7960–7964, (2004).

[107] S. Auer, M. A. Miller, S. V. Krivov, C. M. Dobson, M. Karplus, and M. Vendruscolo. Importance of metastable states in the free energy landscapes of polypeptide chains, *Phys. Rev. Lett.*, **99**, 178104–4, (2007).

[108] S. Auer, C. M. Dobson, and M. Vendruscolo. Characterization of the nucleation barriers for protein aggregation and amyloid formation, *HFSP J.*, **1**, 137–146, (2007).

[109] T. Neuhaus, O. Zimmermann, and U. H. E. Hansmann. Ring polymer simulations with global radius of curvature, *Phys.l Rev. E*, **75**, 051803–10, (2007).

[110] N. Metropolis, A. Rosenbluth, M. Rosenbluth, A. Teller, and E. Teller. Equation of state calculations by fast computing machines, *J. Chem. Phys.*, **21**, 1092, 1087, (1953).

[111] H. M. Berman, J. Westbrook, Z. Feng, G. Gilliland, T. N. Bhat, H. Weissig, I. N. Shindyalov, and P. E. Bourne. The protein data bank, *Nucl. Acids Res.*, **28**, 235–242, (2000).

[112] K. Wolff, M. Vendruscolo, and M. Porto. A stochastic method for the reconstruction of protein structures from one-dimensional structural profiles., *Gene*, **422**, 47–51, (2008).

[113] F. Teichert and M. Porto. Vectorial representation of single- and multi-domain protein folds, *Eur. Phys. J. B*, **54**, 131–136, (2006).

[114] K. Wolff, M. Vendruscolo, and M. Porto. Stochastic reconstruction of protein structures from effective connectivity profiles, *PMC Biophysics*, **1**, 5, (2008).

[115] G. M. D. Mori, G. Colombo, and C. Micheletti. Study of the villin headpiece folding dynamics by combining coarse-grained Monte Carlo evolution and all-atom molecular dynamics, *Proteins*, **58**, 459–471, (2005).

[116] U. Hobohm, M. Scharf, R. Schneider, and C. Sander. Selection of representative protein data sets, *Protein Sci.*, **1**, 409–417, (1992).

[117] W. Kabsch and C. Sander. Dictionary of protein secondary structure: pattern recognition of hydrogen-bonded and geometrical features, *Biopolymers*, **22**, 2577–2637, (1983).

[118] D. Frishman and P. Argos. Knowledge-Based secondary structure assignment, *Proteins*, **23**, 566–579, (1995).

[119] G. Graziano, F. Catanzano, A. Riccio, and G. Barone. A reassessment of the molecular origin of cold denaturation, *J. Biochem.*, **122**, 395–401, (1997).

[120] N. Calosci, C. N. Chi, B. Richter, C. Camilloni, Å. Engström, L. Eklund, C. Travaglini-Allocatelli, S. Gianni, M. Vendruscolo, and P. Jemth. Comparison of successive transition states for folding reveals alternative early folding pathways of two homologous proteins, *Proc. Natl. Acad. Sci. U.S.A.*, **105**, 19241–19246, (2008).

[121] T. K. Chiu, J. Kubelka, R. Herbst-Irmer, W. A. Eaton, J. Hofrichter, and D. R. Davies. High-resolution x-ray crystal structures of the villin headpiece subdomain, an ultrafast folding protein, *Proc. Natl. Acad. Sci. U.S.A.*, **102**, 7517–7522, (2005).

[122] Y. Chen, F. Ding, and N. Dokholyan. Fidelity of the protein structure reconstruction from inter-residue proximity constraints, *J. Phys. Chem. B*, **111**, 7432–7438, (2007).

[123] R. Sathyapriya, J. M. Duarte, H. Stehr, I. Filippis, and M. Lappe. Defining an essence of structure determining residue contacts in proteins, *PLoS Comput. Biol.*, **5**, e1000584, (2009).

[124] E. Alm and D. Baker. Prediction of protein-folding mechanisms from free-energy landscapes derived from native structures, *Proc. Natl. Acad. Sci. U.S.A.*, **96**, 11305–11310, (1999).

[125] A. Caflisch. Network and graph analyses of folding free energy surfaces, *Curr. Opin. Struct. Biol.*, **16**, 71–78, (2006).

[126] R. B. Best and G. Hummer. Reaction coordinates and rates from transition paths, *Proc. Natl. Acad. Sci. U.S.A.*, **102**, 6732–6737, (2005).

[127] G. M. Torrie and J. P. Valleau. Monte Carlo free energy estimates using non-Boltzmann sampling: Application to the sub-critical Lennard-Jones fluid, *Chem. Phys. Lett.*, **28**, 578–581, (1974).

[128] N. A. Denesyuk and J. D. Weeks. Equilibrium and nonequilibrium effects in the collapse of a model polypeptide, *Phys. Rev. Lett.*, **102**, 108101–4, (2009).

[129] A. Barducci, G. Bussi, and M. Parrinello. Well-Tempered metadynamics: A smoothly converging and tunable Free-Energy method, *Phys. Rev. Lett.*, **100**, 020603–4, (2008).

[130] H. Lei, C. Wu, H. Liu, and Y. Duan. Folding free-energy landscape of villin headpiece subdomain from molecular dynamics simulations, *Proc. Natl. Acad. Sci. U.S.A.*, **104**, 4925–4930, (2007).

[131] H. Lei and Y. Duan. Two-stage folding of HP-35 from *ab initio* simulations, *J. Mol. Biol.*, **370**, 196–206, (2007).

[132] D. P. Landau and K. Binder, *A Guide to Monte Carlo Simulations in Statistical Physics*. Cambridge: Cambridge University Press, (2000).

[133] U. H. E. Hansmann and L. T. Wille. Global optimization by energy landscape paving, *Phys. Rev. Lett.*, **88**, 068105, (2002).

[134] C. J. McKnight, P. T. Matsudaira, and P. S. Kim. NMR structure of the 35-residue villin headpiece subdomain, *Nat. Struct. Biol.*, **4**, 180–184, (1997).

[135] J. Kubelka, E. R. Henry, T. Cellmer, J. Hofrichter, and W. A. Eaton. Chemical, physical, and theoretical kinetics of an ultrafast folding protein, *Proc. Natl. Acad. Sci. U.S.A.*, **105**, 18655–18662, (2008).

[136] G. Jayachandran, V. Vishal, and V. S. Pande. Using massively parallel simulation and Markovian models to study protein folding: Examining the dynamics of the villin headpiece, *J. Chem. Phys.*, **124**, 164902–12, (2006).

[137] S. Trebst, M. Troyer, and U. H. E. Hansmann. Optimized parallel tempering simulations of proteins, *J. Chem. Phys.*, **124**, 174903–6, (2006).

[138] W. Vermeulen, P. Vanhaesebrouck, M. V. Troys, M. Verschueren, F. Fant, M. Goethals, C. Ampe, J. C. Martins, and F. A. M. Borremans. Solution structures of the C-terminal headpiece subdomains of human villin and advillin, evaluation of headpiece F-actin-binding requirements, *Protein Sci.*, **13**, 1276–1287, (2004).

[139] R. H. Havlin and R. Tycko. Probing site-specific conformational distributions in protein folding with solid-state NMR, *Proc. Natl. Acad. Sci. U.S.A.*, **102**, 3284–3289, (2005).

[140] Y. Tang, D. J. Rigotti, R. Fairman, and D. P. Raleigh. Peptide models provide evidence for significant structure in the denatured state of a rapidly folding protein: The villin headpiece subdomain, *Biochemistry*, **43**, 3264–3272, (2004).

[141] G. Jayachandran, V. Vishal, A. E. García, and V. S. Pande. Local structure formation in simulations of two small proteins, *J. Struct. Biol.*, **157**, 491–499, (2007).

[142] M. Karplus and D. L. Weaver. Diffusion-collision model for protein folding, *Biopolymers*, **18**, 1421–1437, (1979).

[143] A. Fernandez, M. Shen, A. Colubri, T. R. Sosnick, R. S. Berry, and K. F. Freed. Large-Scale context in protein folding: Villin headpiece, *Biochemistry*, **42**, 664–671, (2003).

[144] J. Kubelka, W. A. Eaton, and J. Hofrichter. Experimental tests of villin subdomain folding simulations, *J. Mol. Biol.*, **329**, 625–630, (2003).

[145] S. V. Krivov and M. Karplus. Hidden complexity of free energy surfaces for peptide (protein) folding, *Proc. Natl. Acad. Sci. U.S.A.*, **101**, 14766–14770, (2004).

[146] S. Piana and A. Laio. Advillin folding takes place on a hypersurface of small dimensionality, *Phys. Rev. Lett.*, **101**, 208101-4, (2008).

[147] T. Herges and W. Wenzel. Free-Energy landscape of the villin headpiece in an all-atom force field, *Structure*, **13**, 661–668, (2005).

[148] J. S. Yang, S. Wallin, and E. I. Shakhnovich. Universality and diversity of folding mechanics for three-helix bundle proteins, *Proc. Natl. Acad. Sci. U.S.A.*, **105**, 895–900, (2008).

[149] M. A. Robien, G. M. Clore, J. G. Omichinski, R. N. Perham, E. Appella, K. Sakaguchi, and A. M. Gronenborn. Three-dimensional solution structure of the E3-binding domain of the dihydrolipoamide succinyltransferase core from the 2-oxoglutarate dehydrogenase multienzyme complex of Escherichia coli, *Biochemistry*, **31**, 3463–3471, (1992).

[150] M. Gruebele. Downhill protein folding: evolution meets physics, *C. R. Biol.*, **328**, 701–712, (2005).

[151] M. Knott and H. S. Chan. Criteria for downhill protein folding: Calorimetry, chevron plot, kinetic relaxation, and single-molecule radius of gyration in chain models with subdued degrees of cooperativity, *Proteins*, **65**, 373–391, (2006).

[152] G. Zuo, J. Wang, and W. Wang. Folding with downhill behavior and low cooperativity of proteins, *Proteins*, **63**, 165–173, (2006).

[153] S. S. Cho, P. Weinkam, and P. G. Wolynes. Origins of barriers and barrierless folding in BBL, *Proc. Natl. Acad. Sci. U.S.A.*, **105**, 118–123, (2008).

[154] M. Jäger, H. Nguyen, J. C. Crane, J. W. Kelly, and M. Gruebele. The folding mechanism of a β-sheet: the WW domain, *J. Mol. Biol.*, **311**, 373–393, (2001).

[155] M. Sudol. Structure and function of the WW domain, *Prog. Biophys. Mol. Bio.*, **65**, 113–132, (1996).

[156] N. Ferguson, J. Berriman, M. Petrovich, T. D. Sharpe, J. T. Finch, and A. R. Fersht. Rapid amyloid fiber formation from the fast-folding WW domain FBP28, *Proc. Natl. Acad. Sci. U.S.A.*, **100**, 9814–9819, (2003).

[157] Y. Mu, L. Nordenskiöld, and J. P. Tam. Folding, misfolding, and amyloid protofibril formation of WW domain FBP28, *Biophys. J.*, **90**, 3983–3992, (2006).

[158] J. C. Crane, E. K. Koepf, J. W. Kelly, and M. Gruebele. Mapping the transition state of the WW domain β-sheet, *J. Mol. Biol.*, **298**, 283–292, (2000).

[159] H. Nguyen, M. Jäger, A. Moretto, M. Gruebele, and J. W. Kelly. Tuning the free-energy landscape of a WW domain by temperature, mutation, and truncation, *Proc. Natl. Acad. Sci. U.S.A.*, **100**, 3948–3953, (2003).

[160] F. Noé, C. Schütte, E. Vanden-Eijnden, L. Reich, and T. R. Weikl. Constructing the equilibrium ensemble of folding pathways from short off-equilibrium simulations, *Proc. Natl. Acad. Sci. U.S.A.*, **106**, 19011–19016, (2009).

[161] P. L. Freddolino, F. Liu, M. Gruebele, and K. Schulten. Ten-Microsecond molecular dynamics simulation of a Fast-Folding WW domain, *Biophys. J.*, **94**, L75–L77, (2008).

[162] Z. Qin, J. Ervin, E. Larios, M. Gruebele, and H. Kihara. Formation of a compact structured ensemble without fluorescence signature early during ubiquitin folding, *J. Phys. Chem. B*, **106**, 13040–13046, (2002).

[163] A. Baba and T. Komatsuzaki. Construction of effective free energy landscape from single-molecule time series, *Proc. Natl. Acad. Sci. U.S.A.*, **104**, 19297–19302, (2007).

[164] G. G. Maisuradze, A. Liwo, and H. A. Scheraga. How adequate are one- and two-dimensional free energy landscapes for protein folding dynamics?, *Phys. Rev. Lett.*, **102**, 238102–4, (2009).

[165] P. Das, S. Matysiak, and C. Clementi. Balancing energy and entropy: A minimalist model for the characterization of protein folding landscapes, *Proc. Natl. Acad. Sci. U.S.A.*, **102**, 10141–10146, (2005).

Die VDM Verlagsservicegesellschaft sucht für wissenschaftliche Verlage abgeschlossene und herausragende

Dissertationen, Habilitationen, Diplomarbeiten, Master Theses, Magisterarbeiten usw.

für die kostenlose Publikation als Fachbuch.

Sie verfügen über eine Arbeit, die hohen inhaltlichen und formalen Ansprüchen genügt, und haben Interesse an einer honorarvergüteten Publikation?

Dann senden Sie bitte erste Informationen über sich und Ihre Arbeit per Email an *info@vdm-vsg.de*.

Sie erhalten kurzfristig unser Feedback!

VDM Verlagsservicegesellschaft mbH
Dudweiler Landstr. 99　　　　　Telefon　+49 681 3720 174
D - 66123 Saarbrücken　　　　　Fax　　　+49 681 3720 1749
www.vdm-vsg.de

Die VDM Verlagsservicegesellschaft mbH vertritt

Printed by Books on Demand GmbH, Norderstedt / Germany